# Dispelling Chemical Engineering Myths

*Forethoughts*

*It ain't so much the things we don't know that get us in trouble. It's the things we know that ain't so.*

Artemus Ward

*Cultures are slow to die; when they do, they bequeath large deposits of custom and value to their successors, and sometimes they survive long after their more self-conscious members suppose them to have vanished.*

Irving Howe, *World of our Fathers*

*Time strips our illusions of their hue*
*And one by one in turn, some grand mistake*
*Casts off its bright skin yearly like the snake.*

Byron, *Don Juan*

*Superstitions take a long time to die—almost as long as new ideas take to be born.*
W.S. Sykes *Essays on the First Hundred*
*Years of Anaesthaesia*

*Tread softly because you tread on my dreams.*
W.B. Yeats, *He Wishes for the Clothes of Heaven*

# Dispelling Chemical Engineering Myths

## Third Edition

**Trevor Kletz**
*Loughborough University*
*Loughborough, England*

Taylor & Francis
*Publishers since 1798*

| USA | Publishing Office: | Taylor & Francis<br>1101 Vermont Avenue, NW, Suite 200<br>Washington, DC 20005-521<br>Tel: (202) 289-2174<br>Fax: (202) 289-3665 |
| | Distribution Center: | Taylor & Francis<br>1900 Frost Road, Suite 101<br>Bristol, PA 19007-1598<br>Tel: (215) 785-5800<br>Fax: (215) 785-5515 |
| UK | | Taylor & Francis Ltd.<br>1 Gunpowder Square<br>London EC4A 3DE<br>Tel: 0171 583 0490<br>Fax: 0171 583 0581 |

**DISPELLING CHEMICAL ENGINEERING MYTHS, Third Edition**

1 2 3 4 5 6 7 8 9 0    BRBR    9 8 7 6

This book was set in Times Roman by Princeton Editorial Associates. The editors were Christine Williams and Carol Edwards. Cover design by Michelle Fleitz. Printing and binding by Braun-Brumfield, Inc.

A CIP catalog record for this book is available from the British Library.

∞ The paper in this publication meets the requirements of the ANSI Standard Z39.48-1984 (Permanence of Paper)

**Library of Congress Cataloging-in-Publication Data**
Kletz, Trevor A.
  Dispelling chemical engineering myths / Trevor Kletz.—3rd ed.
    p.  cm.
  Rev. ed. of: Improving chemical engineering practices. 2nd ed. c1990.
Includes index.
  1. Chemical engineering.   I. Kletz, Trevor A. Improving chemical engineering
practices.  II. Title.
TP155.K544    1996
660—dc20                                                          96-7090
                                                                        CIP
ISBN 1-56032-438-4 (case)

# Contents

# Preface

During the 38 years I spent in the chemical industry, as a manager and, for the last 14, as a safety adviser, I came to realize that many of my colleagues accepted uncritically a number of statements of doubtful accuracy. I therefore thought that it would be useful to set down these myths, as I have called them, and to dispell them, partly because they have led to accidents and wrong decisions and partly to encourage a skeptical approach, particularly among younger engineers.

The first, and longest, part of the book covers myths about technology; the second part, myths about management; and the third part, myths about toxicology and the environment. A short fourth part, a sort of epilogue, reminds us that the chemical industry is not unique, that myths are a feature we find in all walks of life and that all received wisdom should be looked at critically from time to time.

The book was intended primarily for all who work in the chemical industry and for all chemical engineers, including teachers and students. However, as reviewers of the first two editions pointed out, it will also interest all engineers and all who work in industry, including the growing proportion of managers and administrators, and all those who are willing to look critically at received wisdom. The book may help us to avoid accidents and wrong decisions and use resources more effectively.

The first edition was called *Myths of the Chemical Industry, or 44 Things a Chemical Engineer Ought NOT to Know.* For the second edition the title was changed to *Improving Chemical Industry Practices—A New Look at Old Myths of the Chemical Industry,* as the publisher thought that the original title might lead to the book's being indexed and classified incorrectly. The new title was, however, rather dull, and for this third edition it has been changed again.

For the second (first U.S.) edition, I added another 16 myths to the original 44, making a total of 60 and replaced some words and phrases by their U.S. equivalents. For example, I changed road tanker to tank truck and manager to supervisor. (In the United Kingdom, supervisor is another name for a foreman.) For your reference, there is a glossary of U.S.-UK chemical engineering terms included at the back of the book.

For this third edition I have added 8 myths on technology, 7 on management, and an entirely new section of 16 on toxicology and the environment, making a total of 91. I have also revised some of the original myths and updated the references.

As the book does not form a continuous narrative, it may be better read in installments than at one sitting. The White Queen in *Through the Looking Glass* believed in six impossible things before breakfast. This book may help you disbelieve in one or two.

Some of the myths are now met less often than when I first described them. I have nevertheless left them in the book, as we still come across them from time to time (see the quotation from Irving Howe at the front of the book), and under stress, we often go back to old habits of thought. Managers, for example, who have been on courses advocating consultation and persuasion instead of orders, may revert under stress.

I am sure there are many other myths besides those I have described, and I shall be grateful for any others that readers can send me.

Thanks are due to the many colleagues, past and present, who have suggested ideas for this book or commented on the drafts, particularly Professor F. P. Lees, Mr. K. Palmer, and Dr. A. K. Barbour, and to the Science and Engineering Research Council (first edition) and the Leverhulme Trust (2nd edition) for financial support.

The information in the book is given in good faith but without warranty, and readers should satisfy themselves that the recommendations are appropriate for their specific circumstances. Myth M3 shows what can result when we apply methods of calculation outside the ranges in which they are valid.

# Introduction

Many people believe that scientists and technologists are entirely rational in their beliefs and actions, at least when acting in a professional capacity. Ordinary, uneducated people may believe in myths and act accordingly, but not engineers. A historian of the 18th century has written

> . . . by accustoming landowners and merchants to think in objective terms about how their resources might be employed most profitably, the ethos of capitalism led them to devalue (although not abandon) purely emotional criteria in evaluating men and measures in other spheres. Learning to think in a hard-headed way about profits helped to relax the hold of older irrational feelings as guiding principles in public policy.[1]

Though scientists and technologists may be less prone to a belief in myths than other people, they have not totally abandoned myths, as I shall show. Many half-truths are half believed, and they are not mere whims, like a belief in fairies. They affect our actions. I shall describe many accidents and wrong decisions that were, in part, caused by a belief in myths.

Myths have several characteristics:

1.  They are not completely untrue; there is usually a measure of truth in them, but they are not completely or literally true either.

2. They were often more true in the past than they are now.
3. They are deeply ingrained. When our reasons for believing in them are shown to be invalid, we look for other reasons, or continue to act as if the myths were still true. Once in the mind, they are there to stay.

Readers may like to consider the extent to which these features of myths apply in other walks of life. Here we shall be concerned only with some myths of the process industries, particularly the chemical industry, deeply ingrained beliefs that are not wholly true. Few, if any, people believe all the myths listed below; most believe— or half believe—some of them. This is therefore a work of iconoclasm; an attempt to destroy some myths and to encourage a skeptical approach. Too many engineers seem to accept the received wisdom and see themselves as practitioners of established techniques. Perhaps I can sow seeds of doubt. Most of what we have learned is good and sound, but there is some chaff among the wheat.

A final feature of myths is that often we take them for granted and never think critically about them. Once we write them down, as in the headings to the following sections, we begin to have doubts.

In the following pages the myths are divided into myths about technology, myths about management, and myths about toxicology and the environment. Many of them previously appeared in references 2 through 7 and have a bias toward loss prevention and process safety.

*References*

1. Endelman, T.E. 1979. *The Jews of Georgian England,* p. 35. Philadelphia, PA: Jewish Publication Society of America.
2. Kletz, T.A. 1974. In *Loss Prevention and Safety Promotion in the Process Industries.* Proceedings of the First International Symposium on Loss Prevention and Safety Promotion in the Process Industries, p. 309. Amsterdam: Elsevier.
3. Kletz, T.A. 1976. *J Hazardous Materials* 1(2):165.
4. Kletz, T.A. 1977. *J Hazardous Materials* 2(1):1.
5. Kletz, T.A. 1986. *Chem Engineer* 431:29.
6. Kletz, T.A. 1987. *Chem Engineer* 443:44.
7. Kletz, T.A. 1989. *Chem Engineer* 467:31.

# Myths About Technology

It is important that students bring a certain ragamuffin, barefoot irreverence to their studies; they are not here to worship what is known, but to question it.

J. Bronowski, *The Ascent of Man*

In 1966 there was a fire, described in Myth 2 in a refinery at Feyzin, France. Several large pressure vessels containing liquefied petroleum gas (LPG) were exposed to the flames and ruptured, causing extensive loss of life. The company for which I worked handled significant quantities of LPG and other liquefied flammable gases, and it was one of my responsibilities to see that the lessons learned were passed on to my colleagues and, once learned, not forgotten. I discussed the fire with groups of 12-20 people from various departments, in the manner described in Myth M5. I described the fire briefly and illustrated it with slides. I then asked the group to say why the fire had occurred and what action should be taken to prevent similar incidents in the future.

The response was usually the same: if the vessels had burst, the relief valves must have been faulty, too small, or blocked in some way. I assured my audience that this was not the case. After awhile someone suggested that the metal had been softened by the fire and had lost its strength. This had in fact happened. Below the liquid level the boiling liquid kept the metal cool, but above the liquid level the flames heated the metal so much that it lost its strength and the vessels burst even though the relief valves had kept the pressure in the vessels at or below the design pressure.

I realized that many, perhaps most, engineers believed that a relief valve, properly sized, designed, and maintained would always prevent a vessel from bursting and that protection against overheating was not necessary. In fact, a relief valve will prevent a vessel from bursting at and somewhat above design temperature but not at higher temperatures. The belief was a myth, a widely accepted belief that is not always true. Once people thought critically, they realized that the belief could not be true. Everyone knows that metal loses its strength when it gets hot, but they had never thought about it before; they just accepted the belief uncritically.

Having seen how fallible engineers could be, I looked out for other myths and was surprised how many I found. I described them in a series of papers and, ultimately, in this book.

## MYTH 1

### Pressure vessels must be fitted with relief valves (or rupture discs)

Many people are under the impression that, by law, all pressure vessels must be fitted with relief valves (or rupture discs), and in some countries this is so. In the United Kingdom, however, the law requires only certain pressure vessels to be fitted with relief devices: the Factories Act (Sections 32, 35, and 36) requires steam boilers and receivers and air receivers to be fitted with relief valves, and the Chemical Works Regulations (1922) require certain other vessels to be fitted with them. These regulations, however, apply to only a small portion of the chemical industry, to those processes listed in the regulations, all of which were in use in 1922. In some countries the laws are more restrictive, and in others, including the United States, less so. Nevertheless, whatever the law, for many years it has been the universal practice to fit relief valves on all pressure vessels, and it is usually taken for granted that they will be fitted. They are required by the recognized pressure vessel codes, and it is usually assumed that they are necessary to fulfill the requirement of the UK Factories Act (Section 176) that equipment shall be operated and maintained "in an efficient state, in an efficient working order and in good repair" and the requirement of the Health and Safety at Work Act (Section 6) that any article for use at work must, so far as is reasonably practicable, be "so designed and constructed as to be safe and without risks to health."

However, as plants and equipment expand and increase, the cost of providing relief valves rises. The valves are not so expensive in themselves, but if flammable materials are being handled, the associated blowdown and flare systems are expensive and give rise to complaints about the noise and light. If toxic materials are being handled, the relief valve discharge may have to pass through a scrubbing system.

Is it possible to use, instead of a relief valve, a pressure switch that will detect a rise in pressure and isolate the source of pressure (referred to below as an instrumented protective system or trip)? For example, on a distillation column the relief valve is usually sized on the assumption that the full heat input to the base continues when reflux is lost. Could the rise in pressure be used to isolate the source of heat? If so, only a much smaller relief valve will be needed to cope with other sources of overpressure. If a single pressure switch and motor valve are considered insufficiently reliable, then they can be duplicated.

The reaction of many people is that this suggestion is unacceptable because "instruments are unreliable." "Fit an instrument if you wish," many engineers will say, "so that the relief valve does not lift so often, but let me fit a relief valve as a 'last resort'."

The answer to this comment is that although instruments are not 100% reliable, relief valves are not 100% reliable either, and it is possible to design an instrument system with a reliability as good as or better than that of a relief valve.[1,2] Table 1.1 shows typical figures for a trip system and a relief valve assuming a demand rate of once a year (that is, if there was no relief valve or trip, the vessel would be overpressurized once per year) and assuming that the trip develops a fail-danger fault once in 2 years and a relief valve develops a fail-danger fault once in 100 years.[3]

The duplicated trip is actually *safer* than a relief valve and, if cheaper, could be used instead, provided that the staff concerned understands the need for regular testing and has the necessary facilities. In fact, a fully duplicated trip system is much more reliable than a relief valve, and complete duplication is often unnecessary. Because of the uncertainties in the figures and differences in the mode of failure (a relief valve that fails to operate at the set pressure may operate at a higher pressure, but this is not true of a trip), I suggest

**TABLE 1.1**  Hazard rates for various protective systems

| Protective system | Hazard rate (i.e., the frequency with which the design pressure is exceeded) |
|---|---|
| No trip or relief valve | Once a year (assumed) |
| Simple trip, | |
| annual testing | Once in 5 years |
| monthly testing | Once in 48 years |
| weekly testing | Once in 200 years |
| Relief valve, | |
| annual testing | Once in 200 years |
| Duplicated trip, | |
| monthly testing | Once in 1728 years |

that a trip system used instead of a relief valve should have a reliability 10 times greater than the figures quoted above for relief valves.

What about vessels that must be fitted with relief valves by law? Can protective systems be used instead?

As far as steam boilers and air receivers are concerned, the law in the United Kingdom is absolute—relief valves must be fitted; for steam receivers the law is less definite. However, we would not often want to fit protective devices instead of relief valves because steam and air can be discharged safely into the atmosphere. As far as vessels covered by the Chemical Works Regulations are concerned, Regulation 5 states that a relief valve must be fitted to "Every still and every closed vessel . . . in which the pressure is liable to rise to a dangerous degree." It could be argued that if a reliable trip is installed, then the pressure is not liable to rise to a dangerous degree.

Similar arguments can be applied to the pressure vessel codes. The Appendix to BS 5500 states

> J.1.1 Every vessel shall be protected from excessive pressure or vacuum . . . except as provided for in J.1.2.
> J.1.2 When the source of pressure (or temperature) is external to the vessel and is under such positive control that the pressure (or temperature) cannot exceed the design pressure (or temperature), a pressure (or temperature) protective device need not be provided.

A reliable trip, properly maintained, keeps the pressure under positive control.

The predecessors of BS 5500 (BS 1500 and BS 1515) contained similar wording (though without the references to temperature).

The corresponding U.S. code (ASME Boiler and Pressure Vessel Code, Section VIII) does not contain similar wording and requires all pressure vessels to be fitted with pressure relief devices (Section UG-125). However, it seems that the letter of the code can be satisfied if a small pressure relief device is fitted and a reliable trip is installed to keep the pressure under control. Section UG-133 states that

> The aggregate capacity of the pressure relieving devices . . . shall be sufficient to carry off the maximum quantity that can be generated or supplied to the attached equipment.

Presumably, in calculating the "maximum quantity," allowance can be made for the effects of a reliable trip system.

*Examples of the Use of Protective Systems in Place of Relief Valves*

The following are some examples of situations in which instrumented protective systems have been, or might be, used in place of relief valves.

1. It is sometimes possible for a compressor, handling flammable gas, to suck a vacuum in the suction catchpot. Relief valves are sometimes installed to prevent this from occurring, but sucking air into the vessel with a consequent risk of explosion could be as dangerous as collapsing the vessel. On some systems of this type, protective systems have been installed to detect a low pressure, trip the compressor, and close a valve in the suction line before the pressure gets too low. The systems have been designed so that failure will coincide with a demand and the vessel will be underpressurized once in 2000 years.

2. On a distillation column, as already stated, the relief valve is usually sized on the assumption that the heat input to the reboiler continues when reflux or cooling has been lost. This usually calls for a large relief valve. On a few distillation columns a high-pressure trip has been used to isolate the steam to the reboiler or to isolate the fuel to the reboiler furnace. A small relief valve can then be installed to cope with excess pressure generated in other ways. Care must be taken that the residual heat in a furnace or reboiler is not sufficient to overpressurize the column.

3.  As part of an uprating project, a scrubbing column had to be fitted with an additional steam reboiler and feed vaporizer. A larger relief valve would normally be required. The flare stack and flare header were not able to take any additional load, and their replacement by a larger system would have been very expensive. A trip system was therefore designed to isolate the steam to the reboiler and shut down the hot water pump on the feed vaporizer when the pressure approached the set pressure of the relief valve. The relief valve was left in position to cover fire relief and other requirements.[2]

4.  Flammable gas at high pressure in a pipeline often has to be let down to a lower pressure. Normally, a relief valve is fitted on the low-pressure gas main to prevent it from being overpressurized if the reducing valve fails in the open position. A very large relief valve is required. If there is no convenient flare stack available and no suitable site for one, a high-reliability trip system can be installed to isolate the high-pressure main if the reducing valve fails and the downstream pressure rises.

5.  Protective systems would not normally be used in place of relief valves on steam duty because steam can be discharged safely into the atmosphere. However, large condensing turbines are a special case because they need very large relief valves to cover accidental isolation of the condensate lines or loss of cooling water. Failure to provide these valves has caused many accidents.[4] Protective systems that isolate the steam when the pressure in the turbine rises should therefore be considered.

6.  In a particular plant, hydraulic overpressure of a vessel is prevented by a trip system that isolates the feed. In addition, several alarms have to be ignored before there is any demand on the trip system.

7.  In oxidation plants, air (or oxygen) and hydrocarbons have to be mixed, and sometimes the plants have to operate close to the explosive limit. Installation of a relief device large enough and sufficiently quick acting to prevent an explosion from occurring is usually impracticable. Some companies, therefore, install blast walls around the equipment. These do not prevent damage to the reactors but merely minimize the consequences to people and other equipment. (Sometimes

they do not even do that; see Myth 18.) Stewart[5] has described an oxidation plant that operates very close to the explosive limit and is fitted with a protective system of exceptional reliability, similar to those installed on nuclear power stations. Fifty trip initiators, all of them triplicated, detect the approach of dangerous conditions and isolate the oxygen supply by means of a high-reliability shutdown system.

8. In semibatch reactions (in which one component is placed in the reactor and the other added gradually as the reaction proceeds) a common cause of runaway reactions is loss of mixing. A high-integrity protective system can be used to detect this and stop addition of the second reactant. In fully batch processes it is less easy to dispense with relief valves; a protective system can increase cooling or add a reaction stopper, but this may not be sufficient; each case needs individual consideration.[6]

*Compromise Solutions*

If, despite the foregoing information, the pull of tradition is too strong, then a number of compromise solutions are possible.

1. Fit a full-size relief valve, but let it discharge into the atmosphere, though this would not normally be allowed, and rely on a trip system to prevent it from lifting or, to be more precise, to greatly reduce the chance that it will lift. Each case must be considered on its merits, but in many cases, relief valves discharging flammable or toxic gas into the atmosphere should not lift more than once in several hundred years, and relief valves discharging hazardous liquids into the atmosphere should not lift more than once in several thousand years.

2. When several relief valves discharge into the same flare header, the flare header can be sized to take the flow from the largest relief valve (or largest pair), and protective systems can be used to ensure that the probability that more than one (or two) relief valve(s) will lift at the same time is acceptably small. For example, on one plant a number of distillation column relief valves discharge into a flare header that is sized to take the rate from the largest only. All the columns could be over-pressurized simultaneously if there was a power failure or cooling

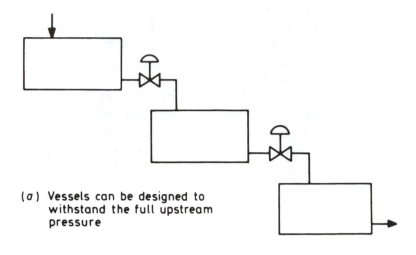

(*a*) **Vessels can be designed to withstand the full upstream pressure**

(*b*) **Vessel can be designed to withstand pump closed-head delivery pressure**

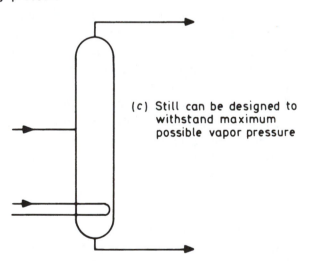

(*c*) **Still can be designed to withstand maximum possible vapor pressure**

**FIGURE 1.1**   Relief valves (but not fire relief valves) can be avoided by using stronger vessels.

water failure, but instrumented protective systems isolate the heat supply to the base of each column and ensure that the chance is small that more than one relief valve will lift.

3. Occasionally one hears of a vessel in which the pressure is controlled by an instrumented protective system but an undersized relief valve is installed as a sort of token observance of the myth. There is no sense in this (though it may be sensible to install a trip to avoid the use of a large relief valve, and a small relief valve to guard against other sources of overpressure).

*Stronger Vessels*

Another way of avoiding the use of relief valves is to use stronger vessels. If a distillation column can withstand the maximum pressure developed when the reflux is lost but heat input to the base continues, then a relief valve is unnecessary [Figure 1. 1(c)]. This may not be economical on a large column but may be on a small one.

Figure 1.1(a) shows a series of vessels with a pressure drop between them. Relief valves are needed on the later vessels in case the letdown valves fail to open and subject the vessels to the full upstream pressure. If all the vessels are made strong enough to withstand this pressure, then relief valves are unnecessary. Similarly, in Figure 1.1(b), if the vessel can withstand the pump delivery pressure, then a relief valve is not needed. In all these cases, if the equipment contains flammable materials, then a fire relief valve will be needed.

*References*

1. Kletz, T.A. 1974. *Chem Processing* Sept p. 7.
2. Kletz, T.A., and Lawley, H.G. 1975. *Chem Engineering* 12 May, p. 81.
3. Kletz, T.A. 1992. *Hazop and Hazan—Identifying and Assessing Process Industry Hazards,* 3rd ed., Sections 3.5.2–3.5.6. Rugby, England: Institution of Chemical Engineers.
4. Naughton, D.A. 1968. *Loss Prevention* 2:54.
5. Stewart, R.M. 1971. In *Major Loss Prevention in the Process Industries.* Symposium Series No 34, p. 99. Rugby, England: Institution of Chemical Engineers.
6. Marrs, G.P., Lees, F.P., Barton, J., and Scilly, N. 1989. *Chem Eng Res. Design* 64(4):381.

# MYTH 2

## A relief valve, properly designed and maintained, will prevent a vessel from bursting; special protection against overtemperature is unnecessary or at least a luxury

A relief valve will prevent a vessel from bursting if it is at, or near, its design temperature but not if the vessel gets too hot. Operating staffs sometimes (though less often now than in the past) fail to realize that if a vessel gets too hot it may burst at or below the design pressure of the relief valve and that the relief valve provides no protection. The protection provided against overtemperature, if any, is usually primitive compared with the protection against overpressure provided by a relief valve.

Vessels can get too hot in several ways, and these are considered separately.

### Vessels Heated by Electricity

Vessels that are heated by steam or hot oil usually cannot get too hot because they are normally designed to withstand the temperature of the heating medium, which is, of course, the maximum temperature attainable. Electric heaters, on the other hand, will overheat if the flow through them stops or gets too low. They are not self-regulating in the same way as steam or oil heaters. Heat input continues whatever the temperature. Protection against high temperature is therefore as necessary as protection against high pressure and, in this case, can be provided just as easily.

Nevertheless, several electric heaters have been installed (or originally designed) without any high-temperature trip operated by the shell temperature. (They may have been fitted with high-temperature trips to protect the heating elements, operated by the temperature of the heating elements, but the set points were too high to protect the shell.) This shows that the lack of protection against high temperature is due not to technical difficulties but to a "blind spot," a failure to realize that high temperature is dangerous. Even when a high-temperature trip is installed, it is usually easy to disarm it (i.e., render it inoperable), whereas disarming a relief valve is usually not possible.

### Internally Insulated Vessels

Internally insulated vessels, such as reactors, are similar to vessels heated by electricity. The internal temperature is often higher than the shell will stand. Any deterioration of the insulation may overheat the vessel. Yet how often do we provide a high-temperature alarm or trip on the vessel wall?

### Furnace Tubes

Vessels are usually tested to 1.5 times their design pressure and will usually withstand several times the design pressure before they burst. Often the vessel does not burst, but a flanged joint stretches and relieves the pressure. In contrast, furnace tubes will often withstand only a 5% or 10% increase in their absolute operating temperature without bursting. Nevertheless, the precautions taken to prevent bursting are usually primitive. Indirect protection may be obtained by a low-flow trip, but these are not always fitted, are often easy to disarm, and provide no protection against flame impingement. Direct protection can be obtained by tube skin temperature measurements, but these are usually considered insufficiently reliable to operate trips and are usually used only as warnings to the operator. Often we do not know what is the hottest part of the tube and so cannot fit the tube skin thermo-couple in the right place.

It is admittedly difficult to protect furnace tubes against over-temperature with anything like the degree of reliability that we take for granted on overpressure protection. Our ignorance in this field is the result of the relative lack of effort expended in the past. If we had put as much effort into devising ways of protecting furnace tubes against excessive temperature as we have put into relief valve design, then the problems might have been solved. As it is, we failed to recognize that excessive temperature presents a hazard in many ways more serious than excessive pressure.

### Vessels Exposed to Fire

In 1966 at Feyzin, France, a leak of propane from a storage sphere caught fire. The fire brigade was advised to use the available water to cool neighboring spheres to prevent the spread of fire. It was assumed that the sphere on fire could be left to itself because the relief valve would take care of it. After 1½ hours, the sphere burst, killing 18 people and injuring many more.[1-6] The refinery staff had failed to appreciate that if a vessel gets too hot it will burst at or below the relief valve set pressure and that the relief valve will not prevent this from occurring. Initially, the vessel was filled, and the boiling liquid removed the heat. As the level fell, the unwetted upper portion of the vessel became heated.

It was suggested that the incident had occurred because the relief valve was too small. It was, in fact, somewhat smaller than some companies would have used, but this was not the cause of the incident.

Vessels that are exposed to a fire can be protected by

water cooling
fireproofing the vessel
reducing the pressure
sloping the ground

Water cooling has been the main line of defense adopted in the past, and the fire services are now usually fully aware of the need to apply cooling water as soon as possible.

Vessels have burst in as little as 10 minutes, so water must be applied quickly. Fixed equipment is therefore often necessary— either monitors that direct a large stream of water on the vessel or sprays that deliver a smaller quantity of water (10 $L/m^2/min$ or 0.2 $gal/ft^2/min$) over the whole surface.

Fireproof insulation is available as an immediate barrier to heat input and does not have to be commissioned like water. Alternatively, vessels can be covered with clean sand or gravel. Some of the covering should be removed every few years, so that the outsides of the vessels can be inspected.

Sloping the ground ensures that any liquid that does not burn immediately runs off to one side. It must not be allowed to run onto the next plant. A collection pit may be necessary.

In addition, it should be possible to lower the pressure in vessels exposed to fire. Often this can be done through existing process lines. Sometimes a relief valve bypass can be fitted and operated remotely. It is possible to obtain a combined relief valve and motor valve that can be lifted by remote operation but that also functions as a normal relief valve. Vapor depressurizing should be provided on all vessels operating at a gauge pressure above 2 bars (30 psi), and the depressurizing valve should be sized so that the pressure falls to 25% of design in 10 minutes, or 30 minutes if the vessel is insulated.

These methods of protection are illustrated in Figure 2.1, and more details can be found in reference 7.

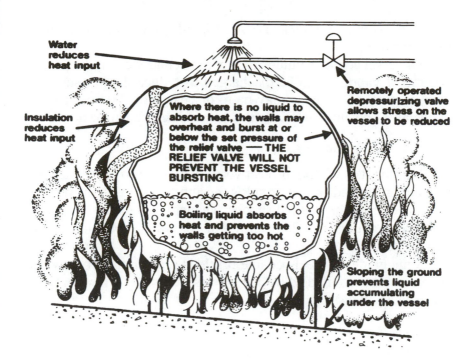

**FIGURE 2.1**   How to protect pressure vessels from fire.

Although all four precautions are recommended, there is some trade-off between them. Thus if insulation is fitted, the water rate can be reduced by two-thirds and the water can be poured over the vessel—it does not have to be sprayed at every part of the surface. Also, the depressurizing valve (and relief valve) can be smaller. If vessels are covered with clean sand or gravel, water cooling is not needed, and the relief valve need not be sized for exposure to fire.

As mentioned above, it is now more widely realized than in the past that relief valves will not protect a vessel that gets too hot. I have discussed the Feyzin fire (or BLEVE—*b*oiling *l*iquid *e*xpanding *v*apor *e*xplosion—as it is often called) on many occasions with groups of operating and design engineers. In the early 1970s the groups would argue that because the vessel burst, there must have been something wrong with the relief valve. Nowadays they usually realize quickly that the metal got too hot.

The British standard for the design of unfired pressure vessels, BS 5500, first issued in 1976, requires vessels to be protected against overtemperature as well as overpressure (see quotation in Myth 1).

Its predecessors, BS 1500 and BS 1515, did not include the reference to temperature.

If a vessel contains liquids of low molecular weight, then a large volume of vapor has to be discharged to reduce the pressure. The quickest way of doing so may be to discharge the liquid through a drain valve,[8] but this has disadvantages, as discussed in the next myth.

### Vessels That Get Too Cold

Relief valves will not protect vessels that get too cold. They may become brittle and fail, at or below design pressure, as the result of inherent stresses or external shocks. The leak and explosion at Beek in the Netherlands in 1975 occurred in this way.[9] Protection by trips may be necessary.

Some grades of steel become brittle at temperatures below 5°C, and vessels have failed while being pressure tested with water below 5°C (see Myth 44).

### References

1. *The Engineer,* 25 March 1966, p. 475.
2. *Paris Match,* No 875, 15 Jan 1966.
3. *Fire,* Special Supplement, Feb 1966.
4. *Petroleum Times,* 21 Jan 1966, p. 132.
5. Lagadec, P. 1980. *Major Technological Risk,* p. 176. Oxford, England: Pergamon Press.
6. Kletz, T.A. 1994. *What Went Wrong?—Case Histories of Process Plant Disasters,* 3rd ed., Section 8.1. Houston, TX: Gulf Publishing.
7. Kletz, T.A. 1977. *Hydrocarbon Processing,* 56(8):98. (Reprinted in *Fire Protection Manual for Hydrocarbon Processing Plants,* ed. C.H. Vervalin, Vol. 2, p. 291. Houston, TX: Gulf Publishing.)
8. Mahgerefteh, H., Giri, N., and Wong, A. 1993. *Chem Engineer* 548(12).
9. van Eijnatten, A.L.M. 1977. *Chem Engineering Prog* 73(9):69.

## MYTH 3

### If a vessel is exposed to fire, it should be emptied as quickly as possible

A vessel exposed to fire should be emptied quickly if a leak from the vessel is feeding the fire—draining the vessel to a safe place will remove the source of fuel and extinguish the fire.

However, if a vessel is being heated by a fire and the fuel is not coming from the vessel, it is usually safer to leave the inventory in the vessel. The boiling liquid will remove the heat. If the vessel is emptied, the metal will soon overheat, the vessel will be damaged, and if it is still under pressure, it will burst violently.[1]

A domestic kettle is safe on a gas stove so long as it contains water. If it boils dry, it is quickly damaged.

As the result of a change in the method of operation, a distillation column was no longer required, and it stood empty in the middle of a unit. If a fire had occurred, this column would have collapsed before the other vessels. I found it difficult to convince the staff that they should fill it with water, although someone suggested that it would be safer to use it for the storage of a high-boiling, though flammable, liquid than to leave it empty. In the end it was dismantled.

If a large number of joints are exposed to fire—particularly certain types of high-pressure joints such as lens ring joints—it may be safer to empty the vessel. It is, of course, necessary to have somewhere to put the contents, and some high-pressure plants are provided with blowdown tanks for this purpose.

It is also desirable to empty vessels containing materials that decompose explosively when the temperature rises.

*Reference*

1. Klaassen, P. 1971. In *Major Loss Prevention in the Process Industries.* Symposium Series No 34, p. 111. Rugby, England: Institution of Chemical Engineers.

## MYTH 4

### Using a vessel designed for a higher pressure is safer than using one designed for the operating pressure

At first sight it seems obvious that a stronger vessel must be safer. Although it is less likely to fail, if it does fail, the damage is greater.

Suppose a vessel is needed for the storage of a liquid at or near atmospheric pressure. A low-pressure tank would normally be provided, but sometimes a redundant pressure vessel is available.

Suppose the vessel is subjected to an excessive pressure because an internal coil bursts, gas is blown into the vessel, the vessel is overfilled, or an internal explosion occurs. The relief valve *should* be designed to guard against the first three but may not be. The low-pressure tank will probably have been designed for a pressure of 8 inches water gauge (4 kPa), and the roof will probably lift off at a pressure of 24 inches water gauge (12 kPa) but the contents will not spill. The pressure vessel *may* withstand the pressure, but if it does not, bits and the contents may fly in all directions.

Suppose the vessel is exposed to fire. The metal is softened and loses its strength (see Myth 2). The low-pressure tank bursts at a low pressure with little consequential damage. The pressure vessel, if not adequately vented, will burst at a higher pressure, again scattering the contents and bits of the vessel in all directions.

When vessels are used that are designed for a higher pressure than is necessary, their relief valves should be set just above operating pressure; they should not be set at the vessel's design pressure.

The methyl isocyanate at Bhopal was stored in a pressure vessel rather than a low-pressure tank so that it could be transferred by nitrogen pressure. When a runaway reaction occurred, the tank was distorted but did not burst. If it had been designed for a lower pressure, it would have burst and the vapor would have been discharged nearer the ground. It would not have spread so far, and fewer people would have been killed.[1]

### An Incident

A tank truck hit a pipe leading to a storage tank, and the pipe broke off inside the bund. The truck's engine ignited the spillage and started a bund fire that destroyed or damaged 21 tanks.

A 100-m$^3$ vertical pressure vessel designed for a gauge pressure of 0.3 bar (5 psi), but used as a storage tank and vented into the atmosphere, contained "methyl ethoxol" (the monomethyl ether of ethylene glycol). Its flash point is 40°C, so it is not flammable at ordinary temperatures, but the fire heated the liquid to its flash point and then ignited the vapor coming out of the vent. The fire flashed back into the vessel, and an explosion occurred. The vessel came apart at the bottom seam, and most of it took off like a rocket, with flames coming out of the base[2,3] (Figure 4.1).

**FIGURE 4.1**  A storage tank goes into orbit.

*Other Examples of Stronger Containment Increasing the Hazard*

Certain explosives are wetted with water or alcohols to suppress their explosive properties. They are packed in strong containers to prevent evaporation of the diluent, but this increases the pressure developed by any explosion that does occur. It might be better to use less volatile diluents and weaker containers.[4]

At first sight, plastic containers seem unsuitable for explosive solids or liquids. However, in a fire the plastic soon melts or burns, and the contents then burn as a pool fire. Confined in a metal drum, the contents might explode, and the pressure would cause the drum to burst violently.[5]

When Guy Fawkes attempted to blow up the Houses of Parliament in 1605, he laid stones and crowbars on top of the barrels of gunpowder. The weight, he knew, would increase the pressure developed by the explosion.[6]

Writing with nuclear reactors in mind, Mosey points out that the classical method of limiting the behavior of a machine is to set a detector close to the upper limit of operation. This is fine when the machine is operating near the limit but not when it is operating at low power. Under these conditions, by the time the upper limit is reached, the power may be increasing so fast that the shutdown system cannot operate quickly enough.[7] It is easier to stop a runaway horse or vehicle when its speed is low than when it is high.

Just as stronger may not be better, so bigger is not always better (see Myth 36).

*References*

1. Kletz, T.A. 1994. *Learning from Accidents,* 2nd ed., Chap. 10. Oxford, England: Butterworth-Heinemann.
2. *Loss Prevention* 1973, 7:119.
3. Manufacturing Chemists Association. 1975. *Case Histories of Accidents in the Chemical Industry,* Vol. 4, Item 1887. Washington, DC: Manufacturing Chemists Association.
4. Merrifield, R., and Roberts, T.A. 1991. *Hazards XI—New Directions in Process Safety.* Symposium Series No 124, p. 209. Rugby, England: Institution of Chemical Engineers.
5. An answer to a question in the discussion in ref. 4, p. 218.
6. Bowen, C.D. 1957. *The Lion and the Throne,* p. 208. London: Hamilton.
7. Mosey, D. 1990. *Reactor Accidents,* p. 17. London: Butterworth Scientific.

# MYTH 5

**It is bad practice (and illegal) to fit a block valve below a relief valve, but operators must be free to disarm trips that protect vessels from the effects of high or low temperatures, high or low levels, high concentrations of dangerous materials, and so on**

To deal with the legalities first, in the United Kingdom, if a vessel must be provided with a relief valve (see Myth 1), then a block (isolation) valve must not be fitted below it, unless two or more relief valves are provided and the block valves are interlocked. (The law is not very clear but is usually assumed to mean this.)

Many companies extend this legal requirement to all relief valves. Operators, they consider, make mistakes and may leave a block valve shut after a relief valve has been changed. This is a perfectly reasonable view. The same companies, however, may have an almost casual attitude toward trips and alarms that protect equipment from the effects of high or low temperatures, high or low levels, dangerous concentrations of corrosive or explosive materials, and so on. Such trips may be easy to disarm, may be disarmed by operators on their own initiative, and may be left disarmed after trip testing, and there may be no system for indicating whether or not they are disarmed.

Why is it that disarming a relief valve fills us with horror, but disarming a trip, which may protect against a more dangerous condition than overpressure, is often routine? There is no logical technical reason for the difference; we have unthinkingly absorbed the folklore of the industry.

If we accept that there is no logic in the present situation, what should we do? Should we treat trips like relief valves, or relief valves like trips?

We should use the methods of hazard analysis (quantitative risk assessment)[1,2] to estimate the probability and extent of a dangerous incident; that is, we should estimate

1. the probability that the relief valve will be called on to lift while it is isolated or the probability that the trip will be called on to operate while it is disarmed, and
2. the severity of the consequences.

We should then compare the results of (1) and (2) with a target or criterion. The results of such an analysis will show that we can safely treat some relief valves with less sanctity than in the past and fit block valves below them (see example below) and that many companies should treat trips with greater respect than in the past.

*Occasions When Relief Valves Can Safely Be Fitted with Block Valves*

Relief valves can be safely fitted with block valves on the following occasions:

1. when the maximum pressure that can be developed with the relief valve isolated will not take the equipment above its test pressure (or, more precisely, when the worst combination of pressure and temperature will produce stresses in the equipment less than those produced by the pressure test)
2. when the relief valve is protecting a long pipeline containing a liquid below its flash point (see example in section below) and one of the line isolation valves can be locked open
3. when it is possible to lock open an alternative vent before isolating the relief valve
4. when equipment isolation valves can be locked open so that the equipment is protected by another relief valve of adequate size.
5. when the demand rate on the relief valve is reduced to less than 0.1/year (once in 10 years) by an instrumented protective system that is regularly tested and maintained

*The Control of Trip Disarming*

It may be necessary to disarm trips (and alarms) from time to time or alter their set points, but this should be done in a controlled manner. Operators should not be allowed to disarm or alter trips (or alarms) at will. The essential features of a control system are as follows.

1. Disarming or alteration of trip or alarm settings must be approved in writing by an authorized person, normally a supervisor (plant manager in the United Kingdom) or senior foreman. Artificers should not alter trip or alarm settings without written authorization.
2. If a trip is disarmed, this should be signaled in some way, for example, by a light on the panel.

3. Trips and alarms should be tested regularly, say, every month primarily to detect failures but also to show if the trip (or alarm) has been disarmed or its setting altered.

4. If trips have to be disarmed for start-up (e.g., a low-flow trip), then they should rearm themselves automatically after a short period of time (several minutes to an hour, depending on circumstances).

### An Example of the Application of Hazard Analysis: The Isolation of Relief Valves

Relief valves on many interplant pipelines are difficult to release for testing. Should we therefore fit block valves beneath them, or is this too risky? Instead, should we go to the expense of twinned valves and interlocked block valves? As a result of the following calculation, it was decided to install block valves below relief valves on pipelines handling liquids *below* their flash points.

The relief valves are installed to prevent the pipelines from being overpressurized if they are isolated full of liquid and the temperature then rises. If a relief valve is left isolated in error, then the pipeline will leak at a pair of flanges.

The following was assumed:

1. There are 100 line relief valves in the area covered by a single operator (or four shift operators).
2. Each relief valve is isolated for inspection once every 2 years.
3. Once in every 100 occasions the procedure breaks down, one relief valve is left isolated, and this is not detected until the next inspection (i.e., on average, at any time one relief valve is isolated).
4. The "demand rate" on each relief valve is once per year.
5. The operator is always near enough to one pipeline to be exposed to spray from a flange (i.e., for 1% of the time the operator is near the pipeline with the isolated relief valve).
6. The chance of the operator being hit by the spray is 1 in 10.
7. The chance of the operator being injured sufficiently seriously to result in a lost-time accident (LTA) is 1 in 5.

These assumptions are all based on the judgment of experienced people, and they all err on the side of safety. If we accept these assumptions, then it follows that

1. One pipeline will be overpressurized and will leak at a flange once in 2 years. (The period between tests is 2 years, and there are two demands in this period, but after the first demand, the fact that the relief valve is isolated will be discovered.)
2. An operator will be exposed once in 200 years.
3. An operator will be hit by spray once in 2000 years.
4. An operator will be injured once in 10,000 years or once in $8.75 \times 10^7$ hours, equivalent to an LTA frequency rate of 0.001 (in $10^5$ hours).

Most companies consider an LTA rate of 0.5 as good and 0.1 as excellent.

A hazard that increases the LTA rate by 0.001 is so small that we should not allocate resources to reduce it further when there are so many potentially bigger risks to be dealt with first. Relief valves on pipelines carrying oils below their flash points may therefore be fitted with block valves.

Two points regarding this analysis should be noted:

1. Most examples of hazard analysis deal with fatal accidents. The method can also be applied to LTAs.
2. We have assumed that an administrative procedure will be set up to prevent block valves under relief valves from being left isolated, but nevertheless we have assumed that sooner or later, this procedure will break down. We have thus tried to estimate the frequency of breakdowns.

*References*

1. Kletz, T.A. 1992. *Hazop and Hazan—Identifying and Assessing Process Industry Hazards,* 3rd ed. Rugby, England: Institution of Chemical Engineers.
2. Lees, F.P. 1996. *Loss Prevention in the Process Industries,* 2nd ed., Chap. 9. Oxford, England: Butterworth-Heinemann.

## MYTH 6

**Designers can assume that operators will do what they are asked to do, provided they are properly trained and instructed and the task is within their mental and physical powers**

This myth is believed more by designers than by those who operate plants.

Reports from all industries (and road accident reports) show that more than half the accidents that occur, sometimes 80% or 90%, are due to human failing. If only people would take more care, or would follow instructions, then we would have fewer accidents.

However, if people have not taken care or followed instructions in the past, we cannot assume that they will change their ways in the future. Engineers should accept people as they find them and design accordingly, and leave the changing of human nature to those more qualified to do so (who, judging by the results achieved in the past few thousand years, have not been very successful in doing so).

There is a story about a man who went into a tailor's shop for a ready-made suit. He tried on most of the stock without finding one that fit. Finally, in exasperation, the tailor said, "I'm sorry, sir, I can't fit you. You're the wrong shape."

There are many reasons why people fail to do what they are asked, but in the individual context, three reasons predominate:

1. a moment's aberration
2. poor training
3. lack of supervision

### A Moment's Aberration

Well-trained, well-motivated people, physically and mentally capable of following directions, make occasional mistakes. They know what they should do, and want to do it, but occasionally forget. For example, they forget to open a valve, or they open the wrong valve. Exhortation, punishment, or further training will have no effect. We must either accept an occasional error or change the work situation (that is, the plant or method of operation), so that errors are less likely. We should try to remove or reduce opportunities for error.

Note that errors occur not in spite of the fact that people are well trained, but because they are well trained. Routine activities are delegated to the lower level of the brain and are not continuously monitored by the conscious mind. We would never get through the day if everything required our full attention. When the normal flow or pattern of actions is interrupted, errors may occur.

For example, during the start-up of a unit, an operator forgot to open a valve, and an explosion occurred. The unit and a similar one had between them been started up successfully 6000 times before the explosion occurred.[1] Nevertheless, one explosion in 6000 is too many. An interlock should have been installed, so that the operator could not proceed with the start-up until the valve was open.

In some compressors it is possible to interchange suction and delivery valves, and the result has been damage and leaks. Valves should be designed so that they cannot be interchanged.

Designers should not, of course, assume that operators will always make errors. They should try to estimate the probability of an error, and then decide, taking the consequences into account, whether to accept the occasional error or modify the work situation. Humans are actually very reliable, but there are hundreds, perhaps thousands, of opportunities for error in the course of a day's work, so we should not be surprised that some errors occur.

Another opportunity open to designers is to allow the operator to recognize and correct any errors made. For example, in entering instructions into a computer, an operator sometimes has to check that the instructions are correct and then press the "enter" key a second time. Unfortunately, operators may soon get into the habit of pressing the "enter" key twice without checking. A better system is for operators to carry out two distinct operations after entering instructions, for example, moving a cursor and then pressing "enter."

Slipups are increased by stress and distraction, and sometimes these can be reduced.

### Poor Training

Many incidents have occurred because operators were not able to diagnose faults. Three Mile Island is the best known example.[2] In other cases, operators have not understood how equipment works or what they were required to do. On a number of occasions, operators have written down on record sheets temperatures or pressures that indicated an approaching hazard, but did nothing about them. Ultimately, a dangerous incident occurred.

Designers should bear in mind that the plants they design will probably be operated by people who are prone to repeat the errors made in the past.

### Lack of Supervision

People carrying out routine tasks may become careless and take short cuts. Managers and supervisors cannot stand over them but should check, from time to time, that procedures are being followed. Unfortunately, this does not always happen.

For example, as the result of corrosion on a plant in which water was electrolyzed, some of the hydrogen entered the oxygen stream, and an explosion occurred. The hydrogen and oxygen were supposed to be analyzed every hour for purity. After the explosion, it was found that when plant conditions changed, the oxygen analyses on the record sheet changed immediately, although it would take an hour for such a change to occur on the plant. Clearly, the analyses were not being carried out and the managers had not noticed and had failed to impress on the operators the importance of regular analyses and the serious consequences that could follow if the oxygen purity fell.[3]

Designers should not assume that the plants they design will be operated by superhumans who never take short cuts or break the rules. They should assume that people will behave as they have in the past (see Myth M20). Their designs need not be foolproof, but they should be able to withstand, without serious failure, the sort of errors that experience shows will continue to occur. To take a very simple example, any equipment installed on the plant is liable to be stood upon in spite of warnings not to do so. Equipment should be made strong enough to stand upon, or be shaped or located so that it cannot be stood upon (Figure 6.1).

For other examples, see Myth 30.

*Examples of Human Error from Other Industries*

1. In the early days of anesthetics, chloroform was often mixed with air and piped to a face mask using the apparatus shown in Figure 6.2, introduced in 1867. If the two pipes were interchanged, liquid chloroform was supplied to the patient. This happened on a number of occasions, with fatal results.

   Redesigning the apparatus so that the two pipes could not be interchanged was easy—all that was needed was a difference in size or type of connection—but persuading doctors to use the new designs was another matter. They did not believe that a professional would be capable of such a simple error. They did not realize that slips and lapses of attention can happen to anyone, regardless of ability, training, or intention. As late as 1928, deaths from the use of the simple apparatus were still being reported.

**FIGURE 6.1** Plant equipment should be strong enough to stand upon—or be shaped or located so that it cannot be stood upon.

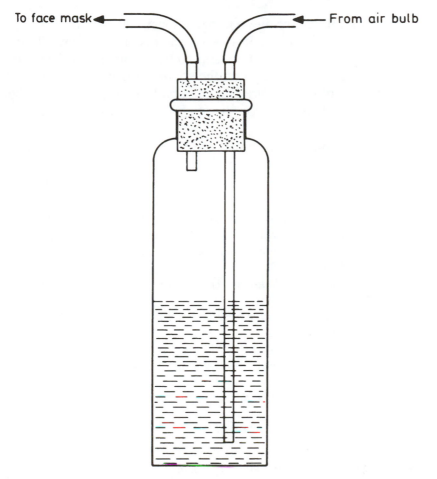

**FIGURE 6.2** Early chloroform dispenser.

Many deaths could have been avoided if anesthetists had used an apparatus redesigned so that errors could not occur.[4]

However, do not assume from this account that doctors are less competent than engineers. In 1989 in Houston, Texas, maintenance workers were dismantling a branch pipe that had been isolated, they thought, from equipment containing hot, high-pressure ethylene gas by means of a valve operated by compressed air. The valve was actually open, and hot ethylene escaped and caught fire, killing 23 people. The valve was open because the two compressed air lines, one to open the valve and one to close it, had been connected the wrong way around.[5]

2.  After a railway bridge had been constructed on the George-
    town Loop line in the Rocky Mountains in 1883, it was found
    that the two supporting towers had been transposed. The north
    tower had been erected on the south bank and *vice versa,* so
    that the entire structure had been built backward. The bridge
    (of rivetted construction) had to be dismantled and rebuilt.[6]

For other examples of human error, see the "Further Reading"
section.

*References*

1.  Vervalin, C.H., ed. 1985. *Fire Protection Manual for Hydrocarbon Processing Plants,* 3rd ed., p. 95. Houston, TX: Gulf Publishing.
2.  Kletz, T.A. 1994. *Learning from Accidents,* 2nd ed., Chap. 11. Oxford, England: Butterworth-Heinemann.
3.  *The Explosion at Laporte Industries Ltd on 5 April 1975.* 1976. London: Her Majesty's Stationery Office.
4.  Sykes, W.S. 1982. *Essays on the First Hundred Years of Anaesthesia,* pp. 3–5. Edinburgh: Churchill Livingstone.
5.  US Department of Labor. 1990. *The Phillips 66 Company Houston Chemical Complex Explosion and Fire 1989.* Washington, DC: US Department of Labor.
6.  Morgan, G. 1976. *Rails Round the Loop,* p. 15. Fort Collins, CO: Centennial Publications.

*Further Reading*

Kletz, T.A. 1991. *An Engineer's View of Human Error,* 2nd ed. Rugby, England: Institution of Chemical Engineers.
Kletz, T.A. 1994. *What Went Wrong—Case Histories of Process Plant Disasters,* 3rd ed. Houston, TX: Gulf Publishing.

# MYTH 7

## Accidents are due to human failing, so we should eliminate the human element when we can

This myth is the opposite to Myth 6. Both assume that accidents
are due to human failing, but Myth 6 assumes that people can be
persuaded to stop making errors, whereas Myth 7 assumes that people
will always make errors, and proposes we try to remove our dependence on them.

To deal with the first part of the myth, "Accidents are due to human failing," as we showed in the discussion of Myth 6, this is true of many accidents, but it is not very helpful to say so. We cannot stop humans from failing, so we should look for other ways of preventing accidents. In any accident investigation, we should look for those causes that we can do something about.

Can we therefore eliminate the human element? Suppose that when an alarm sounds an operator has to go outside, select the right valve out of many, and close it within 10 minutes. He may fail to do so, in a busy control room, on one occasion out of 10 or, in a quiet control room, on one occasion out of 100.

We can eliminate the operator from the loop by arranging for the valve to be closed automatically when the alarm condition is reached. We thus remove our dependence on the operator, but we are now dependent on the people who design, install, test, and maintain the automatic equipment. They also make mistakes. We have not removed our dependence on humans; we have merely transferred it from one human to another (or others). It may be right to do so because the people who design, install, test, and maintain the equipment probably work under less stress and distraction than the operator, and, if so, their error rates will be lower. But do not let us kid ourselves that we have eliminated our dependence on the human factor. Every error is a human error because

Someone has to decide what to do.
Someone has to decide how to do it.
Someone has to do it.

## MYTH 8

**Trips are unreliable: the more trips we install, the more spurious trips we get, so on the whole, it is better to rely on operators**

This myth is heard in America more often than in Europe. American companies on the whole seem less willing than European ones to rely on trips, and more willing to rely on operators. Unfortunately, as shown in the discussion of Myths 6 and 7, the reliability of operators may not be sufficient.

The reliability of a trip can be increased to any desired level by increasing the test frequency or duplicating components. (However, this is true only up to a point. If we increase the test frequency too much, the time that the trip is out of action during testing becomes important. If we duplicate too many components, common mode failures become important.[1]) With this qualification, if we specify the reliability required, the instrument engineer can design a trip (or group of trips) with this requirement.

Duplicating a trip does increase the spurious trip rate, but this can be compensated for by installing a voting system. For example, a duplicated trip might typically give a spurious trip every 0.75 year. A two-out-of-three voting system might have a spurious trip once in 100 years. The cost of the voting system can be justified by the savings that result from the avoidance of spurious trips.[1]

*Note:* In a two-out-of-three voting system there are three measuring instruments. Two must indicate a hazard before the trip operates. Spurious trips—the trip operating when it should not because of a fault in the measuring equipment—therefore happen much less frequently.

*Reference*

1. Kletz, T.A. 1992. *Hazop and Hazan—Identifying and Assessing Process Industry Hazards,* 3rd ed., Sections 3.6.4–3.6.5. Rugby, England: Institution of Chemical Engineers.

## MYTH 9

### If a flammable material has a high flash point, it is safe and will not explode

A material will not explode so long as it is below its flash point. Once gas oil, fuel oil, heat transfer oils, and other liquids with high boiling points are heated above their flash points, they become as dangerous as gasoline.

In addition, a liquid will explode at a temperature as much as 300°C below its flash point if it is in the form of a finely divided mist.[1,2] A liquid will ignite at a temperature more than 200°C below its flash point if it is in the form of a thin film.[3] Even though a bulk liquid is below its flash point, a source of ignition may have sufficient energy to produce local heating and ignition.[4,5] At low pressures, gases and vapors can explode below their normal flash points. For example, a jet fuel of flash point 50°C at atmospheric pressure has a flash point of 20°C at a pressure of 1 psi (7 kPa).[6,7]

Contamination by a small amount of low-flash-point liquid will reduce the flash point considerably. For example, 2% gasoline reduced the flash point of a solvent from 37°C to 15°C (see fact 3 below).

Many fires and explosions have occurred because the following facts, particularly the first, were not known to those concerned.

1. A mixture of phenols was separated in a batch vacuum distillation column. The materials all had flash points above 80°C and were considered safe. When a batch was complete, the vacuum was broken with air. After a number of batches, the steam coil in the boiler was withdrawn for cleaning. On one occasion, as it was being withdrawn, an explosion occurred, and the coil shot out of the boiler, fortunately missing the men on the job.

   The coil had been withdrawn many times before without incident, but on this occasion the maintenance team had started work sooner than usual after the completion of a batch, and the distillation column was still hot, above the flash point of the vapors. The source of ignition was sparks or heat produced by withdrawing the coil.

   In the factory where this occurred, gasoline was also handled. No one would have dreamed of adding air to equipment containing gasoline, nor of opening up the equipment while gasoline was still present, but the phenols were considered safe. Those concerned did not realize that when phenols are above their flash points they are as dangerous as gasoline and should be treated with the same respect.

   After the explosion, the vacuum was broken with nitrogen, and the phenols were removed by adding water and bringing it to a boil before opening up the column to remove the heating coil.

   *Note:* Do *not* add water to a vessel that is above 100°C (see Myth 37).

2. A few years later, in the same factory, repairs had to be carried out to the roof of a tank containing phenol. The staff wished to avoid, if possible, emptying and cleaning the tank, but remembering the previous incident, they emptied the tank as

much as possible, isolated the steam coil, and allowed the few inches of phenol in the base of the tank to set solid. (The melting point of phenol is 41°C.) They then allowed a welder to weld a patch on the roof.

The welder's torch vaporized and ignited some phenol that had sublimed onto the roof, and a mild explosion occurred, lifting the roof but not removing it completely (Figure 9.1). Fortunately, the welder saw some fumes coming out of the vent and left the roof just before the explosion.

A tank that has contained heavy oils or solids can never be made perfectly clean. The atmosphere inside must be made inert with nitrogen or inert foam before welding is allowed. Filling with water can reduce the volume to be inserted.

3. A man was cleaning a moped engine using an apparatus in which a high-flash-point (37°C) solvent was pumped from a drum to a cleaning brush, from which it fell into a tray and drained back into the drum. His clothes caught fire, and although he jumped into a pool of water, he later died from his burns.

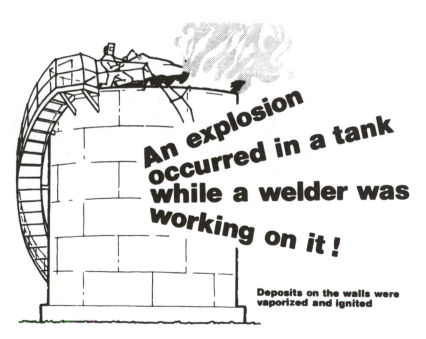

**FIGURE 9.1**   Heavy oils can explode when heated.

It was found that a small amount of gasoline reduced the flash point of the solvent; 2% reduced it to 15°C. The source of ignition was never found with certainty but may have been a spark from the circulating pump motor. It was not flameproof because it was intended for use with a high-flash-point solvent.

Another explosion involving materials of high flash point is illustrated in Figure 13.3.

*References*

1. Eichhorn, J. 1955. *Petroleum Refiner* 5(4):255.
2. Kohlbrand, H.T. 1990. Case history of a deflagration involving an organic solvent/oxygen system below its flash point. Paper presented at AIChE Loss Prevention Symposium, Aug, San Diego, CA.
3. Englund, S.M. 1994. Inherently safer plants—Practical applications. Paper presented at AIChE Summer National Meeting, Aug, Denver, CO.
4. Nettleton, M.A. 1981. *Archivum Combustionis* 1(1/2):131.
5. Plewinsky, B., Wegener, W., and Herrman, K.-P. 1988. *Progr Astronaut Aeronaut* 114.
6. Penner, S.S., and Mullins, B.P. 1959. *Explosions, Detonations, Flammability and Ignition,* p. 119. Oxford, England: Pergamon Press.
7. Ministry of Aviation. 1962. *Report of the Working Party on Aviation Kerosene and Wide-Cut Gasoline.* London: Her Majesty's Stationery Office.

## MYTH 10

### Flammable mixtures are safe and will not catch fire or explode if everything possible has been done to remove known sources of ignition

We are brought up to believe in the fire triangle: air, fuel, and a source of ignition are necessary for a fire or explosion to occur; take one element of the triangle away, and an explosion is impossible.

When flammable gases or vapors are handled on an industrial scale, this view, though theoretically true, is misleading. If flammable gases or vapors are mixed with air in flammable concentrations, then experience shows that sources of ignition are likely to turn up.

In many investigations of fires and explosions, the source of ignition is never found. Sometimes the investigator attributes the ignition to static electricity but without demonstrating the precise way in which static electricity might have been responsible.

The amount of energy required to ignite a flammable mixture can be very small, as little as 0.2 mJ. This is the energy produced when a penny falls 5 mm, though the energy has, of course, to be concentrated into a small time and space as in a spark or speck of hot metal. It is perhaps, therefore, not surprising that we cannot completely eliminate all sources of ignition.

As an alternative to the fire triangle, I suggest

$$AIR + FUEL \rightarrow BANG$$
or
$$AIR + FUEL \rightarrow FIRE$$

What are these mysterious sources of ignition? Sometimes it is static electricity. A steam or gas leak, if it contains liquid droplets or particles of dust, produces static electricity that can accumulate on an ungrounded conductor, such as a piece of wire netting, a scaffold pole, or a tool, and then discharge to ground. Discharges may occur from the cloud itself. In other cases, ignition may be due to traces of pyrophoric material, to traces of catalyst on which reactions leading to local high temperatures may occur, to friction, or to the impact of steel on concrete (but not to the impact of steel on steel; see Myth 12).

The only safe rule is to assume that mixtures of flammable vapors in air in the explosive range will sooner or later catch fire or explode and should never be deliberately permitted, except under carefully defined circumstances where the risk is accepted. One such set of circumstances is in the vapor space of a fixed-roof storage tank containing a flammable conducting liquid such as acetone or methanol. Static electricity is not a serious risk, provided splash filling is not allowed, and experience shows that explosions are very rare. The same is not true of tanks containing flammable hydrocarbons with low flash points, and additional precautions such as nitrogen blanketing are necessary to reduce the risk of explosion to an acceptable level.

Here are just two examples of fires or explosions caused by unusual sources of ignition.

An explosion occurred in a fixed-roof storage tank. The liquid had a high conductivity, so static electricity could be ruled out as the source of ignition. The only source that could be found was frictional heating caused by a taut vibrating wire, which supported a swing arm, rubbing against a pulley that had seized on its bearing and was not free to move. Experiments have shown that steel wires subject to friction can produce glowing filaments of thin wire that cannot ignite methane but might ignite other gases.[1]

A fire in an open tank occurred while a mechanic was tightening a screwed fitting that was leaking.[2] The tank had been emptied but still contained flammable vapor. The tank was made from aluminum, but the fitting—a valve—was made from steel. According to the report, friction between the steel and aluminum caused oxidation of aluminum to aluminum oxide, a reaction that is exothermic. Alternatively, a thermite reaction between aluminum and rusty iron might have occurred.

*References*

1. Kletz, T.A. 1994. *Learning from Accidents,* 2nd ed., Chap. 6. Oxford, England, Butterworth-Heinemann.
2. Manufacturing Chemists Association. 1975. *Case Histories of Accidents in the Chemical Industry,* Vol. 4, Item 2184. Washington, DC: Manufacturing Chemists Association.

*Further Reading*

Bond, J. 1991. *Sources of Ignition.* Oxford, England: Butterworth-Heinemann.


## MYTH 11

### The worst mistake one can make in a plant handling flammable liquids or gases is to introduce a source of ignition

This myth is related to Myth 10. Reports on fires and explosions often show an excessive concern with the source of ignition. If discovered, it is often listed as the "cause." However, since sources of ignition usually turn up once a flammable mixture is formed, the real cause of the fire or explosion is the failure that allowed the flammable mixture to form, either by letting liquid or gas out of the plant or by letting air in. The only sure way of preventing fires and explosions is to keep the fuel inside the plant, and the air out.

Which is the greater crime in a plant handling flammable liquids or gases: to bring in a box of matches or to bring in a bucket (Figure 11.1)? Most people would say that it is more dangerous to bring in the matches, but matches are dangerous only if they are struck

**FIGURE 11.1** Which of these is more dangerous in a plant handling flammable liquids?

**FIGURE 11.2** Buckets are more dangerous than matches.

when a flammable mixture is present, and in a well-run plant that is very rare. If a bucket is used, however, for collecting drips or samples or liquid for cleaning, a flammable mixture is always present above the liquid and may be ignited by a stray source of ignition (Figure 11.2).

I do not suggest that we should allow indiscriminate smoking, welding, and so forth in our plants. Obviously, we must do what we can to remove known sources of ignition, so that those leaks that do occur are less likely to ignite, but this is our second line of defense. The first line is to prevent the formation of a flammable mixture. We should keep buckets out of our plants as conscientiously as we keep out matches.

For an account of a serious explosion that occurred because those concerned thought they had eliminated all sources of ignition and were therefore casual about leaks, see reference 1.

*Reference*

1. Kletz, T.A. 1994. *Learning from Accidents,* 2nd ed., Chap. 4. Oxford, England: Butterworth-Heinemann.


## MYTH 12

### Nonsparking tools should be used in plants that handle flammable liquids or gases

Nonsparking (better described as spark-resistant) tools seem to be regarded as a sort of magic charm to prevent explosions, though a series of reports over 40 years has shown that they have little value in this regard.

The American Petroleum Institute (API) has published a Safety Data Sheet[1] that summarizes these reports. It does not say when the tools were first introduced, but as far back as 1930, a number of engineers were asking if they were really necessary. In 1941 an API report showed that it was very unlikely that petroleum vapor could be ignited by the impact of steel on steel produced by hand, and that power operation is required to produce an incendiary spark. Later investigations confirmed these conclusions. It may be possible to ignite hydrogen, ethylene, acetylene, and carbon disulfide by the impact of steel on steel using hand tools, but we should not let anyone carry out a maintenance job in an explosive atmosphere of hydrogen, ethylene, or anything else.

If a leak is small, a person may be allowed to put his hands, protected by gloves, into a cloud of flammable vapor to harden up a leaking joint. Where hydrogen, ethylene, acetylene, and carbon disulfide are handled, nonsparking hammers should be available for this purpose. It is probably better to use steel wrenches because they will harden up the joint more effectively.

If a flammable cloud extends more than a foot or so from the leak, no attempt should be made to harden up the leaking joint. People should never be asked to put all or most of their bodies into a cloud of flammable vapor to carry out a repair job because the vapor may ignite.

There is no harm in using nonsparking hammers for all purposes, but it is an unnecessary expense. Care must be taken that small particles of grit do not get embedded in the hammers or they will be more dangerous than steel ones.

There is a need for nonsparking tools when explosive substances such as solid explosives are handled. The foregoing discussion applies to mixtures of flammable gas or vapor and air.

*Reference*

1. American Petroleum Institute. 1973. *Safety Data Sheet No PSD 2214.* (Dec). New York: API.

## MYTH 13

**If a combustible gas detector reads zero, then there is no vapor present, and it is safe to introduce a source of ignition**

Combustible gas detectors, both fixed and portable, are some of the most useful instruments we possess and have made an enormous contribution to safety. However, knowledge of their limitations is essential if we are not to be misled. Unfortunately, operators are often too willing to believe a zero reading, perhaps because that is the reading they would like to see.

Some of the causes of incorrect readings are as follows.

1. *The instrument is out of order.* When these instruments fail, they do not always fail safe. Portable instruments should therefore be tested every day or, better still, immediately before each use (Figure 13.1). A useful test material is 30%

**FIGURE 13.1**   Flammable gas detectors are valuable but have limitations.

isopropanol in water because it normally produces a low reading (57% of the lower flammable limit at 18°C) and will thus detect loss of sensitivity. It also prevents damage to the filament of the instrument by repeated exposure to rich mixtures.

2. *The vapors may be absorbed by the sample tube.* For this reason, it is better to use instruments in which the detecting element is placed at the point of test.

3. *The sample tube may be choked* as a result of the swelling caused by absorption of vapors or in other ways.

4. *The element may have been poisoned by exposure,* for example, to halogenated hydrocarbons or silicones. Poisoning by the former is temporary, by the latter permanent.

5. *The substance being detected may form a flammable mixture with air only when hot* and may cool down in the instrument. This is the most common cause of failure to detect explosive

mixtures, and one explosion that occurred as a result will therefore be described in detail.

A furnace tripped out on flame failure as the result of a reduction in fuel oil pressure. The operator closed the two isolation valves and opened the bleed (Figure 13.2).

When the oil supply pressure had been restored, the foreman tested the inside of the furnace with a combustible gas detector. He got no response and therefore inserted a lighted poker; a bang occurred, damaging the brickwork and slightly injuring the foreman.

When the burner went out, it took a few seconds for the solenoid valve to close, and during this time, oil entered the furnace. In addition, the line between the valve and the burner may have drained into the furnace. The vapor from this oil (flash point 65°C) was too heavy to be detected by the combustible gas detector because it condensed out in the sample tube. If a detector in which the detector head is placed at the point of test had been used, the vapor might have condensed out on the sintered metal that surrounds the detector head.

When relighting a hot furnace that burns fuel with a flash point above ambient temperature, we cannot rely on a combustible gas detector to detect a flammable mixture. We should therefore sweep out the furnace for a long enough period of time to be certain that any unburned oil has evaporated. Operators should know the reason for purging, so that they are less likely to rush the purge time.

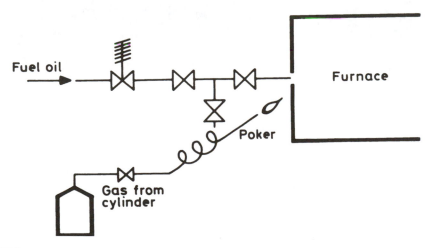

**FIGURE 13.2** A combustible gas detector did not prevent an explosion in this furnace.

**FIGURE 13.3** Heavy oils, which are not detected by gas detectors, can explode when they are heated above their flash points.

It is not a bad rule to say, "If a furnace burning fuel oil trips, have a cup of tea before relighting it." This will give most furnaces time to purge. If the delay is unacceptable, then permanent pilot burners, supplied from a separate fuel supply, may be used.

To keep the purge time as short as possible, the solenoid valve should close quickly, it should be close to the burner, and the line in between should be sloped, so that it does not drain into the furnace.

There is a need for a gas detector that can detect vapor that is explosive when hot but safe at atmospheric temperature, for example, a detector with a heated sample tube.

For many years the furnace on which the explosion occurred and other furnaces burning heavy oils had been tested with detectors that were incapable of detecting the vapor of the fuel. Often we disbelieve instruments; we believe them uncritically when they tell us the answer we want.

Another incident is illustrated in Figure 13.3.


## MYTH 14

**Gasoline engines produce sparks and must not be used in areas in which flammable gases and vapors are handled, but diesel engines are fine**

Although diesel engines do not use sparks, they can ignite flammable mixtures of gas and vapor in air in other ways:

1. The flammable mixture can be sucked into the engine via the air inlet, and can ignite and flash back.
2. Sparks or flames can be produced by the exhaust, or the exhaust manifold or pipe can be hot enough to ignite the mixture.
3. The use of a decompression control can ignite a flammable mixture by exposing it to the hot gases in the cylinder.
4. Ancillary equipment may produce sparks.

In the incident described under Myth 4, a spillage was ignited by a diesel engine, but the precise mechanism is not known.

In another incident, a spillage of 4 tons of hot cyclohexane occurred during a plant shutdown with a diesel-engined vehicle operating in the area. The vapor was sucked into the air inlet, and the engine started to race. The driver tried to stop it by isolating the fuel supply but without success because the fuel was going in with the air. Finally, valve bounce and flashback occurred and ignited the cyclohexane. Four men were killed by the fire.[1]

Proprietary devices are available for protecting diesel engines, so that they can operate safely in areas where leaks may occur. They operate by shutting off the air supply as well as the fuel supply, and can be operated manually or automatically when a leak occurs.[2] It is, of course, also necessary to ensure that flammable vapors cannot be ignited in other ways. A spark arrestor and flame arrestor should be fitted to the exhaust, which should be kept below the auto-ignition temperature of the materials used in the area; the decompression control, if fitted, should be disconnected, so that it cannot be used, and electrical equipment should be protected.[3]

Note that if insulation is used to keep the temperature of the exhaust below the auto-ignition temperature of the materials handled, contact of liquid with the insulation (or even with dust that has settled on the exhaust pipe) may lower the auto-ignition temperature of the liquid.[4]

In June 1989 the press reported[5] that a diesel railway engine had ignited a leak from a liquefied natural gas pipeline near Ulfa, in central Russia. More than 600 people were killed. It is not certain that the diesel engine was the source of ignition. It could have been a spark from the wheels or a cigarette or match thrown out of a window, but the diesel engine seems the most likely source.

*References*

1. Kletz, T.A. 1994. *Learning from Accidents,* 2nd ed., Chap. 5. Oxford, England: Butterworth-Heinemann.
2. *Hazardous Cargo Bulletin,* Jan 1983, p. 22.
3. Oil Companies Materials Association. 1977. *Recommendations for the Protection of Diesel Engines Operating in Hazardous Areas.* Chichester, England: John Wiley & Sons.
4. Stanton, R.C. 1976. *Chem Engineer,* May, p. 391.
5. *Daily Telegraph* (London). 5 and 6 June 1989.

# MYTH 15

## If the pressure on a liquefied gas is reduced, the amount of liquid remaining can be calculated by heat balance

If a liquid is under pressure at a temperature above its atmospheric pressure boiling point and the pressure is reduced, then some of the liquid will flash, and the rest will fall to its boiling point at the new pressure. It is easy to calculate the amount that will flash and the amount that remains. However, the flash is accompanied by the production of a great deal of spray, and experiments have shown that under the right conditions, all the liquid may form a mixture of vapor and spray.[1-3] This phenomenon is well known to anyone who has removed the cap from the radiator of a car while the engine was hot. In this case the spray consists of large droplets that soon fall to the ground. In other cases the spray may be fine.

No one, so far as I know, understands in detail the factors controlling the amount or fineness of the spray. It probably depends on the liquid density, its surface tension, its degree of superheating, and the geometry of the vessel and of the aperture causing the reduction in pressure.

Note that if the vapor is flammable, a fine liquid spray is also flammable and remains flammable below the flash point (see Myth 9).

It is common practice to assume that the amount of spray produced is roughly equal to the amount of vapor produced (unless the vapor amounts to more than half the original liquid, in which case, it is assumed that the rest of the liquid will form spray).

*References*

1. Reed, J.D. 1975. *Proceedings of the First International Symposium on Loss Prevention and Safety Promotion in the Process Industries,* p. 191. Amsterdam: Elsevier.
2. Osigo, C., et al. 1972. In *Proceedings of the First Pacific Chemical Engineering Congress.* Part II, Paper 9-4. Tokyo.
3. Advisory Committee on Major Hazards. 1984. *Third Report: The Control of Major Hazards,* Appendix 5. London: Her Majesty's Stationery Office.

# MYTH 16

## A pressure of 10 pounds is small and will not cause injury.

There is not a misprint in the heading. I wrote it that way because we commonly say the pressure in the vessel is "10 pounds" rather than "10 pounds per square inch."

Unfortunately, this expression leads to a belief that, because 10 pounds is not much, a pressure of 10 pounds is not much. Once we analyze the myth, we see it is wrong, but it is still hard to get the

**If this tank is emptied with the vent shut — the force on the top is the same as if a railway engine was lowered onto it...**

**No wonder it collapses!**

**FIGURE 16.1**   The effect of emptying a tank with the vent shut.

idea out of our heads. Whenever people have been injured or a plant damaged by pressure, surprise is expressed—by technically qualified people as well as operators—that so little pressure could cause so much damage or injury.

For example, some years ago an operator opened the door, 3.5 feet (1.1 m) in diameter, of a steam filter before blowing off the pressure. The operator was crushed by the door against the frame of the filter and was killed instantly. During the investigation, surprise was expressed that such a small pressure (a gauge pressure of 30 psi or 2 bars) could cause such injuries and damage, and a chemical explosion in the filter was suggested. In fact, simple calculation shows that the force acting on the door was 18 tons—and it is not surprising that when the holding bars were released, the door flew open with great violence.

In another incident, a driver opened the manhole on top of a pressure tank truck while there was air inside at a gauge pressure of 10 psi (0.3 bar). He was blown off the top. Surprise was expressed that the pressure was sufficient to cause this.

See Myth 30 for more details of these two incidents.

On several occasions, tank trucks have been emptied with the manhole and vents shut and have collapsed. Surprise has been expressed that the atmospheric pressure is sufficient to cause collapse. In fact, atmospheric pressure acting over the surface area of even a small tank can amount to 200 tons. No one would expect the tank to survive if a railway engine (approximately the same weight) was lowered onto it (Figure 16.1).

The use of SI units, in which force and pressure are measured in entirely different units, may help to avoid confusion. However, it will not prevent accidents from occurring to people who do not understand the nature of vacuum. A laboratory worker saw a colleague, a Ph.D. chemist, drying some material under vacuum in a glass flask, without using a safety shield. When this was pointed out, the chemist protested that he was not using a shield since the vacuum was low, only 29 inches of mercury. It took several minutes of explanation before he realized that the pressure difference between the inside and the outside of the flask was within 3% of that obtainable with a perfect vacuum.

## MYTH 17

### Vessels are strong and can withstand any treatment they are likely to receive

This is a myth believed by many plant operators and even by supervisors whose background is in chemistry rather than engineering.

A pressure vessel originally designed for a gauge pressure of 0.3 bar (5 psi) was used for storing at atmospheric pressure, as liquid, a product with a melting point of 97°C. It was kept hot by a steam coil. The inlet line to the tank was being blown with compressed air at a gauge pressure of 5 bars (75 psi) to prove that it was clear, the usual procedure before filling the tank. The vent on the tank was choked, and the end of the tank was blown off, killing two men who were working nearby.

The operators found it hard to believe that a "blow of air" could burst a steel pressure vessel, and explosion experts had to be brought in to convince them that there had not been a chemical explosion. In fact, the air pressure was far greater than the bursting pressure of the vessel (a gauge pressure of about 1.3 bars or 20 psi). Steel looks stronger than air and, though we may be intellectually convinced that it is not, it is harder to take this for granted.

The vent on the vessel had choked on previous occasions, and the operators had complained about the poor access, which made it difficult for them to see if it was clear and to rod it free if it was not. However, this choking was looked upon, by operators and supervisors, as an inconvenience rather than a hazard, and improvements carried no priority and were never made.[1]

*Reference*

1. Kletz, T.A. 1994. *Learning from Accidents,* 2nd ed., Chap. 7. Oxford, England: Butterworth-Heinemann.

# MYTH 18

## Blast walls provide the ultimate protection against explosions

Despite all our precautions, explosions may occur inside or outside the plant equipment. Certain items of equipment in which the chance of an explosion occurring has been judged to be greater than normal (e.g., certain oxidation reactors) have been surrounded by blast walls to protect people and other equipment from missiles and blast.

The walls may give some people a feeling of security, but that is about all. To withstand the sort of pressures that might be developed, especially in a confined explosion, the walls would have to be so thick that they would cost as much as the plant. If an explosion occurred inside many so-called blast walls, the result would be that people would be hit by a stream of moving concrete instead of a stream of moving air. Even if the wall withstands the shock wave, it may deflect pressure onto people sheltering behind it.

Instead of building blast walls, we would do better to spend our money on reducing the probability that an explosion will occur. The high-integrity protective system described by Stewart[1] is better value for the money than a blast wall because it will prevent explosions in oxidation reactors instead of providing doubtful protection against the consequences.

Equipment that is particularly liable to leak and ignite is sometimes surrounded by fire walls, which prevent the fire from reaching other equipment. These walls serve a useful purpose. They are often spoken of loosely as blast walls, but they are really fire walls.

In small research plants, where the maximum energy release is equivalent to a few kilograms of TNT, construction of reliable blast walls is feasible and is the recommended solution. A high-integrity protective system would be too expensive, and often we do not have the knowledge needed to design one correctly. A method of designing blast walls for small units has been described by High.[2]

### References

1. Stewart, R.M. 1971. *Major Loss Prevention in the Process Industries.* Symposium Series No 34, p. 99. Rugby, England: Institution of Chemical Engineers.
2. High, W.G. 1967. *Chemistry and Industry,* 3 June, p. 899.

## MYTH 19

### Drums are safer than tanks for the storage of flammable liquids because each one contains so little

This myth is (or was) believed by those responsible for writing codes and regulations, as 100 tons of flammable liquid in drums can be stored nearer to a building than the same quantity in a tank.[1] Procedures and codes have to be followed before a new tank is built, but stacks of drums are allowed to grow on spare ground. A few drums are used "temporarily" when stocks are high, and before long they are a permanent part of the scene.

In fact, flammable liquid in drums is much more hazardous than the same quantity in a storage tank. It would be difficult to find a better way of burning liquid quickly than stacking it in thin-walled containers with spaces between them. In a fire in 1973, 90 tons of solvent in drums burned in 9 minutes[2,3] (Figure 19.1). In contrast, if a storage tank is exposed to fire, the roof may lift, but the liquid stays confined in the tank, as does the fire (assuming that the tank is designed so that the roof/wall seam is weaker than the base/wall seam and will give way first).

Storage of aerosol containers is also more hazardous than is generally realized. In 1982 a large single-story warehouse (floor area $110,000 \, m^2$) was destroyed by a fire that started in a stack of cardboard

**FIGURE 19.1**  The scene after a fire in which 90 tons of solvent in drums burned in 9 minutes. Note that the drums were stacked close to buildings that were damaged or destroyed by the fire.

boxes about 5 m high containing aerosol cans. Although the warehouse was protected by sprinklers, they were quite incapable of controlling the fire, and in addition, rocketing aerosol cans spread the fire. The rocketing cans passed through water curtains that were intended to isolate sections of the warehouse.[4]

### References

1. U.K. Health and Safety Executive. 1977 (Jan). *The Storage of Highly Flammable Liquids.* Guidance Note CS/2. London: Her Majesty's Stationery Office.
2. *Fire,* Dec 1973, p. 362.
3. *Fire Prevention,* May 1974, p. 40.
4. Factory Mutual Insurance. 1983. *Record* 60(3):3.

## MYTH 20

### Ton for ton, toxic gases produce more casualties than flammable gases or liquids

The concentrations of toxic vapor that can cause sudden death or injury are much less than lower flammable limits. The spread of toxic vapors cannot be cut short by ignition. Hence we would expect sudden releases of toxic vapors to produce many more casualties than sudden releases of flammable vapors.

In practice, however, this is not the case. For the period 1970–75 the press, including the trade press, reported 34 fires or explosions in the oil and chemical industries (including transport) throughout the world that resulted in five or more fatalities. These amounted to about 600 fatalities in total. I know of only two comparable toxic incidents in the same period causing a total of 28 fatalities. They were (1) an explosion in a refrigerated store that was caused by a leak of natural gas and that resulted in rupture of several ammonia tanks (it is not clear from the report whether the 10 fatalities were caused by the explosion or the ammonia[1]); and (2) a tank burst that killed 18 people.[2]

Simons[3] compared fatalities caused by the transport of flammable and toxic gases in the United States:

> During the period 1931 to 1961 37 persons (non-workers) were killed in LP-Gas flash fires and explosions from accidents involving tank trucks. This is an average of 1.23 fatalities per year. . . . For the years since 1961, an exact tally has not been made, but the annual average is believed to be in the range 1 to 2 fatalities per year.

In addition, an unknown number of people were killed in accidents involving LP-gas tank cars.

In contrast, during the same period, five people were killed in the United States as the result of accidents involving tank trucks and tank cars of chlorine. So far as is known, the transport of other toxic gases caused no fatalities in the United States in this period. Five deaths in 45 years is an average of 0.1 death/year.

The total quantity of flammable flashing liquid in stores, process plants, and transport containers probably exceeds the total quantity of toxic flashing liquid, but this is not sufficient to explain the difference. Marshall[4] estimated Mortality Indices: the average number of people killed by the explosion of a ton of hydrocarbon or the release of a ton of chlorine or ammonia. He finds that the historical record is

| Substance | Mortality index |
| --- | --- |
| Chlorine | 0.30 |
| Ammonia | 0.02 |
| Liquefied flammable gases | 0.60 |
| Unstable substances | 1.50 |

A recent comparison of the $f$-$N$ curves (number of incidents per year versus number of people killed) for highly flammable gases and chlorine plus ammonia shows that, worldwide, incidents killing 10 people occur at about the same rate but that flammable gases are involved in 50 times as many incidents that kill 50 or more people.[5]

Why do sudden releases of toxic vapors kill so many fewer people than expected, despite the serious results that are theoretically possible?

The explanation may be that, while in theory a toxic vapor can spread a very long way and cause many casualties, weather conditions have to be exactly right and this rarely coincides with a leak. In addition, people can often escape; Simons et al.,[6] discussing chlorine, wrote: "People flee instinctively when confronted by the greenish, choking cloud. Flammable gas clouds do not provide such a clear warning of danger." Furthermore, if windows are closed, a toxic gas cloud can pass over houses without causing casualties. On average, therefore, toxic vapors produce fewer casualties than expected.

Another reason why toxic vapors produce fewer casualties than estimated is that the data for the toxicity of chlorine used by many workers are almost certainly wrong by an order of magnitude.[7]

Taylor[8] discussed the reasons why poison gas was not used in World War II and wrote, "Most probably the explanation was the simple calculation that, weight for weight, high explosive was more effective than gas in killing people."

Unfortunately, at Bhopal, in 1984, weather conditions were exactly right, and thousands of people were living in a shanty town close to the factory. The tumbledown dwellings provided no barrier to the entry of the poisonous gas (methyl isocyanate), and more than 2000 people were killed.[9]

*References*

1. *The Times* (London). 19 Sept 1973.
2. *Proceedings of Public Enquiry into Explosion of an Ammonia Storage Tank at Potchefstroom, South Africa on 13 July 1973.* Pretoria, South Africa.
3. Simons, J.A. 1975. Risk assessment method for volatile toxic and flammable materials. Paper presented at Fourth International Symposium on Transport of Hazardous Cargoes by Sea and Inland Waterways, 26–30 October, Jacksonville, FL.
4. Marshall, V.C. 1982. In *Hazardous Materials Spills Handbook,* eds. G.F. Bennett, F.S. Feates, and I. Wilder. New York: McGraw-Hill.
5. Haastrup, P., and Rasmussen, K. 1994. *Process Safety Environ. Protection* 72(B4):205.
6. Simons, J.A., Erdman, R.C., and Naft, B.N. 1974. Risk assessment of large spills of toxic materials. Paper presented at National Conference on Control of Hazardous Material Spills, San Francisco, CA.
7. Institution of Chemical Engineers. 1987. *Chlorine Toxicity Monograph.* Rugby, England: Institution of Chemical Engineers.
8. Taylor, A.J.P. 1965. *English History 1914–1945.* Oxford, England: Clarendon Press, p. 427; and London: Penguin Books, p. 524.
9. Kletz, T.A. 1994. *Learning from Accidents,* 2nd ed., Chap. 10. Oxford, England: Butterworth-Heinemann.

# MYTH 21

## Compressors and distillation columns are complex items of equipment and need our best operators, but furnaces do not

In many companies the new employees are assigned to furnace operation. When they have proven their ability, they may be promoted to the control of the compressors or distillation columns. This implies that there is not much to furnace operation, that anyone can do it.

This attitude has resulted in many expensive furnace tube failures—the result of operators (and sometimes supervisors) not fully understanding the way furnaces work. In particular, operators do not understand the behavior of the metal from which the furnace tubes are made.

Suppose the tubes are designed to operate at 500°C for 100,000 hours (11 years):

If they are operated at 506°C, they will last 6 years.
If they are operated at 550°C, they will last 3 months.
If they are operated at 635°C, they will last 20 hours.

In each case, failure will be by "creep"—the tube will expand, slowly at first and then more quickly, and will finally burst.

If the tubes are operated at 550°C for 6 weeks, they will use up half their creep life during this period and will fail after 5 or 6 years at design temperature instead of 11 years. No matter how gently they are treated, once they have been overheated, they will never forget. A furnace tube has a better memory than an elephant (Figure 21.1).

An elephant has a good memory....

.... BUT A FURNACE TUBE HAS A BETTER ONE !

If you let your furnace tubes run 60° C hotter than design for 6 weeks, you may halve the life of the furnace.

**FIGURE 21.1**   The effect of temperature on the life of a furnace tube.

If a compressor or distillation column has a throughput greater than design, then we can use the extra capacity. However, if we get more out of a furnace by turning up the burners, we may pay for it later.

Other incidents involving furnaces have occurred because operators did not fully understand the principles underlying the lighting-up procedure and were tempted to take short cuts[1,2] (see Myth 13, item 5).

A casual attitude toward furnaces is not new. In the 19th century, boiler explosions were frequent, but engine breakdowns were much less common. This was due "in particular to the care taken in maintenance which even nowadays is in sharp contrast to the frequent neglect of the steam generator."[3]

### References

1. Institution of Chemical Engineers. 1982. *Furnace Fires and Explosions.* Safety Training Package No. 005. Rugby, England: Institution of Chemical Engineers.
2. Kletz, T.A. 1994. *What Went Wrong—Case Histories of Process Plant Disasters,* 3rd ed., Section 10.7. Houston, TX: Gulf Publishing.
3. Eyers, J. 1967. *Maschinen Schaden* 40(1):3.

## MYTH 22

### If we want a piece of equipment repaired, all we need do is point it out to the person who is going to repair it

Unfortunately, experience shows that the repairman goes to get his tools or finish another job, and when he returns, he cuts open the wrong pipeline, unbolts the wrong joint, or dismantles the wrong pump. Chalk marks are no better—they get washed off by rain or there are so many of them from past jobs that the repairman attends to the wrong one.

The recommended method for identifying equipment that is to be repaired is to attach a numbered tag to the equipment at the point of repair. If a pipeline has to be cut, for example, the tag should be attached at the point at which it is to be cut. The tag number should be put on the work permit that is given to the repairman (Figure 22.1). If equipment has a permanent number painted on it or attached to it, this should be used instead of the tag. Permanent numbers should follow a logical sequence (Figure 22.2).

# Tag along with us

## They make the job exact and neat... Please return when job complete.

**Figure 22.1**   Equipment that is to be maintained should be identified by a numbered tag.

There were seven pumps in a row

A fitter was given a permit to do a
job on No.7.  He assumed No.7
was the end one and dismantled it.
Hot oil came out.

The pumps were actually numbered:

Equipment that is given to
maintenance must be labeled.
If there is no permanent label, then
a numbered tag must be tied on.

**FIGURE 22.2**   Permanent labels should follow a logical sequence.

*Further Reading*

Kletz, T.A. Hazards in chemical system maintenance: permits. In *Safety and Accident Prevention in Chemical Operation,* eds. H.H. Fawcett and W.S. Wood, Chap. 36. New York: John Wiley & Sons.

Kletz, T.A. 1994. *What Went Wrong—Case Histories of Process Plant Disasters,* 3rd ed., Chap. 1. Houston, TX: Gulf Publishing.

Kletz, T.A. 1994. In *Handbook of Highly Toxic Materials Handling and Management,* eds. S.S. Grossel and D.A. Crowl, Chap. 11. New York: Marcel Dekker.

# MYTH 23

## We should do all we can to remove hazards

Thirty years ago this statement would have been accepted without hesitation. Safety was a black-and-white affair. If a hazard was recognized, certainly if an accident had occurred, then action had to be taken to prevent the accident from happening again. Since resources are limited, this often resulted in lavish spending to remove hazards that had been brought to our attention, by an accident or in other ways, accompanied by a reluctance to look too hard for other hazards in case we found more than we could deal with.

Now, to a large extent, the chemical industry has come to realize that we should search systematically for hazards, using techniques such as hazard and operability studies,[1,2] and then use a systematic technique, preferably numerical, such as quantitative risk assessment (hazard analysis),[1,3] to decide which hazards should be dealt with immediately and which can be postponed, at least for the time being.

The Canvey Island Reports[4,5] (quantitative estimates of risk to the public from the oil and chemical plants on an island in the Thames estuary, near London) were milestones in the adoption of these techniques. The methods, data, and criteria used in the reports have been criticized, but nevertheless, they demonstrate an acceptance by government that one cannot do everything possible to prevent every conceivable accident and therefore we should try to quantify the risks and compare them with a target or criterion.

In the United Kingdom a series of further reports have confirmed and extended the official commitment to the quantification of risks, where appropriate.[6-8] In some countries the authorities have gone too far by asking for the quantification of every risk and setting unrealistic targets. In the United States, the Occupational Safety and Health

Administration (OSHA) still seems reluctant to admit that we cannot do everything to remove every risk, however trivial or unlikely to occur.

The widespread adoption of these methods in the safety field contrasts with the environment, where there is less acceptance of the fact that we cannot do everything at once and that a rational method of deciding priorities is needed (see the Part 3 of this book).

*References*

1. Kletz, T.A. 1992. *Hazop and Hazan—Identifying and Assessing Process Industry Hazards,* 3rd ed. Rugby, England: Institution of Chemical Engineers.
2. Lees, F.P., 1996. *Loss Prevention in the Process Industries,* 2nd ed., Chap. 8. Oxford, England: Butterworth-Heinemann.
3. Lees, F.P., 1996. *Loss Prevention in the Process Industries,* 2nd ed., Chap. 9. Oxford, England: Butterworth-Heinemann.
4. *Canvey, An Investigation of Potential Hazards from Operations in the Canvey Island/Thurrock Area.* 1978. London: Her Majesty's Stationery Office.
5. *Canvey—A Second Report.* 1981. London: Her Majesty's Stationery Office.
6. Health and Safety Executive. 1989. *Quantified Risk Assessment: Its Input to Decision Making.* London: Her Majesty's Stationery Office.
7. Health and Safety Executive. 1989. *Criteria for Land-Use Planning in the Vicinity of Major Industrial Hazards.* London: Her Majesty's Stationery Office.
8. Health and Safety Executive. 1992. *The Tolerability of Risk from Nuclear Power Stations,* 2nd ed. London: Her Majesty's Stationery Office.

# MYTH 24

## We can remove hazards

The belief that we can remove hazards is related to the previous myth—if we say we should remove all hazards, we imply that we are able to do so. Unfortunately, we cannot always do so, for several reasons:

1. Safety is often approached asymptotically. There is a small chance that a relief valve will fail and a vessel will be overpressurized. This chance can be made even smaller by fitting two full-size relief valves, but the risk of overpressurizing is still not zero because coincident failure is possible.[1]
2. We will never remove all hazards because we shall fail to foresee all of them. We do try to foresee what might go wrong

by the use of techniques such as hazard and operability studies,[1] but the people who use the techniques are not omniscient, and some hazards will be missed. In particular, we may fail to foresee the consequences of modifications (see Myth 25).

3. The third reason we cannot remove all hazards is that humans will make occasional errors and we cannot remove our dependence on humans by installing automatic equipment. All we can do is to transfer our dependence from one human being to another (see Myth 7).

*Reference*

1. Kletz, T.A. 1992. *Hazop and Hazan—Identifying and Assessing Process Industry Hazards,* 3rd ed. Rugby, England: Institution of Chemical Engineers.

## MYTH 25

### The successful worker is the one who gets things done quickly

At one time the "go-getter" who brushed difficulties aside was considered the successful worker, but this is no longer true. When plants or methods of operation are changed, those who rush repent at leisure. Changes to plants and methods of operation often have unforeseen side effects. The successful worker is now the one who says, "I know it's urgent, but we can afford to spend an hour or two looking critically and systematically at the proposed modification." This is quite a cultural change.

Of course, we seek ideas for change as much as ever, but we also want to try to find out all the consequences before we proceed.

The following are some examples of simple changes to plants or methods of working that had unforeseen and unwanted side effects.

1. When storage tanks are exposed to fire, they are kept cool by pouring water over them, from fixed installations or mobile monitors. The water is poured or directed onto the roof and runs down the sides.

    In 1968 a new British standard (BS 2654, Part 3) was introduced. It allowed tanks to be made with thinner walls, provided the walls were reinforced by wind girders. When the first tanks were constructed according to the new design, it was found that the girders prevented the cooling water from running down the sides of the tanks. Special deflection plates had to be fitted to redirect the water onto the walls.

A better solution, if the consequences had been foreseen, would have been to put the girders *inside* the tanks (Figure 25.1).

2. A vent line was arranged as shown in Figure 25.2(a). When repairs were made, a straight-through cock was not available, so a right-angle type was used [Figure 25.2(b)]. When the vent was used, the reaction forces caused it to whip around, as shown in Figure 25.2(c). Fortunately, no one was hurt, though a similar movement of a vent pipe did cause a fatal accident.[1]

If a right-angled cock had to be used, it should have been installed as shown in Figure 25.2(d).

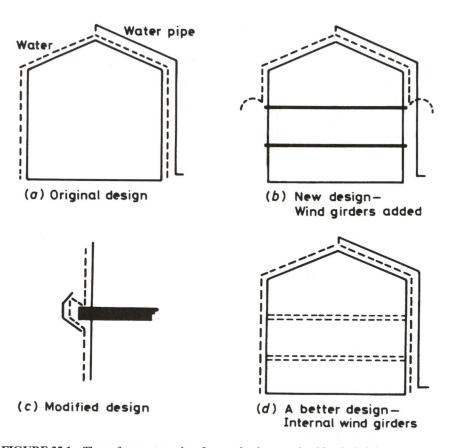

(*a*) Original design

(*b*) New design—
Wind girders added

(*c*) Modified design

(*d*) A better design—
Internal wind girders

**FIGURE 25.1**   The unforeseen results of strengthening a tank with wind girders.

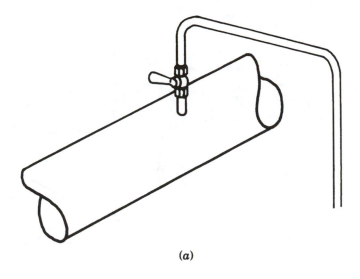

(*a*)

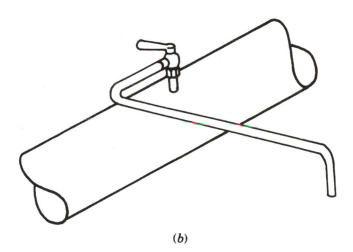

(*b*)

**FIGURE 25.2** (a) Original arrangement of vent pipe. (b) Replacement of straight-through cock with a right-angle type.

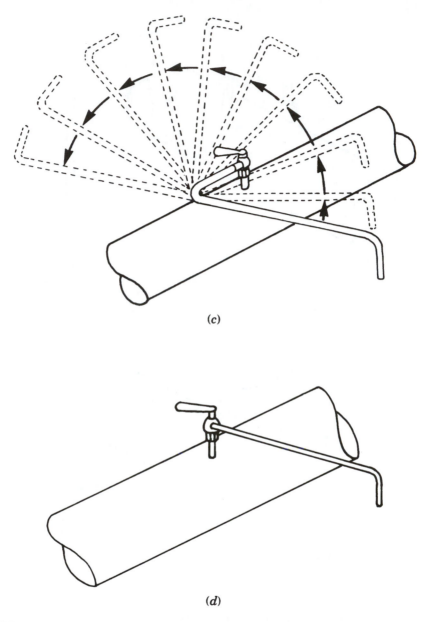

(c)

(d)

**FIGURE 25.2** (*continued*)    (c) Reaction force causing the cock to whip around. (d) Proper installation of a right-angled cock.

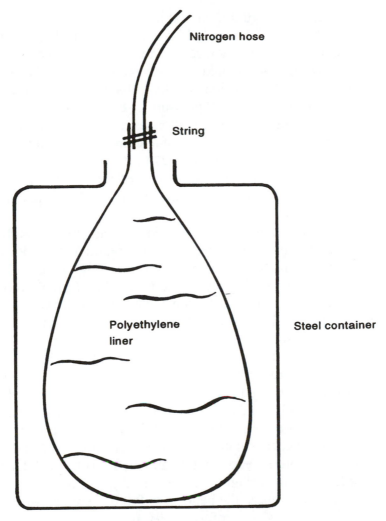

**Nitrogen hose**

**String**

**Polyethylene liner**

**Steel container**

**FIGURE 25.3**   When the polyethylene liner was inflated, the steel container was bowed out.

3.  Large containers, about 60 m³ in volume, had to be filled with powder. A large plastic bag was put into each container and inflated with nitrogen. The normal practice was to put the nitrogen hose inside the neck of the plastic bag, but one day, to save time, an operator tied the neck to the hose with string (Figure 25.3). Fifteen minutes after turning on the nitrogen, there was a loud crack, and the top of the metal container was found to have bowed out by about 3 inches.

The plastic (0.007 inch thick) is not, of course, stronger than the steel (0.064 inch thick), but the plastic bag was larger than the steel container, so the steel gave way first.

The operator should not have connected the hose to the plastic bag without first checking that the container and bag could withstand the full pressure of the nitrogen.

4. The most famous modification ever made in the chemical industry was the installation of a temporary pipe at the Nypro factory at Flixborough in 1974. It failed 2 months later, causing the release of about 50 tons of hot cyclohexane, which mixed with the air and exploded, killing 28 people and destroying the plant.[2-4]

On the plant were six reactors in series; each was slightly lower than the one before, so that the liquid in them flowed by gravity from No. 1 down to No. 6 through short, 28-inch-diameter connecting pipes [see Figure 25.4(a)]. To allow for expansion, each 28-inch pipe contained a bellows.

One of the reactors developed a crack and had to be removed (for the cause, see 5 below). It was replaced by a temporary 20-inch pipe, which had two bends in it to allow for the difference in height. The existing bellows were left in position at each end of the temporary pipe [Figure 25.4(b)].

The design of the pipe and support left much to be desired. The pipe merely rested on scaffolding. Because there was a bellows at each end, it was free to rotate or squirm, and did so when the pressure rose a little above the normal level, though still below the design pressure and the set point of the relief valve. This caused the bellows to fail [Figure 25.4(b)].

There was no professionally qualified engineer on the plant at the time the temporary pipe was built. The men who designed and built it—design is hardly the word because the only drawing was a full-scale sketch in chalk on the workshop floor—did not know how to design large pipes that are required to operate at high temperatures (150°C) and gauge pressures (150 psi or 10 bars) and made no attempt to think through the results of the modification. Very few engineers have the specialized knowledge to design highly stressed piping, but in addition, the engineers at Flixborough did not know that design by experts was necessary or that modifica-

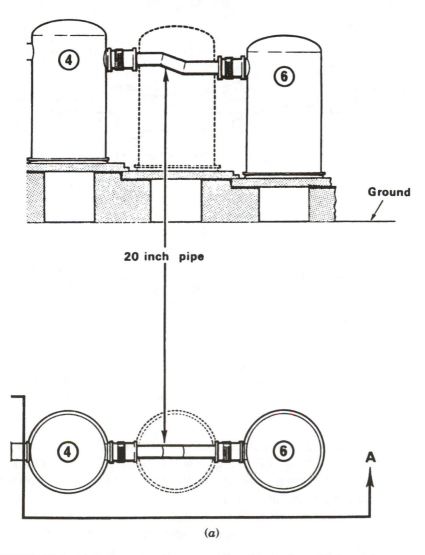

20 inch pipe

Ground

(a)

**FIGURE 25.4**    (a) Arrangement of reactors and temporary pipe at Flixborough.

tions should be probed systematically. They did not know
what they did not know (Figure 25.5).

5. The crack in the Flixborough reactor was itself the result of a
modification. There was a leak from the stirrer gland on the
top of the reactor. To condense the leaking vapor, water was
poured over the top of the reactor. Plant cooling water was
used because it was conveniently available.

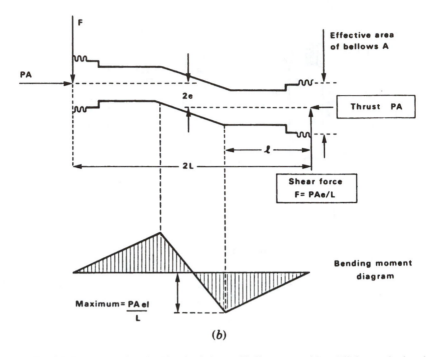

(*b*)

**FIGURE 25.4** (*continued*)    (b) Sketch of pipe and bellows assembly at Flixborough, showing shear forces on bellows and bending moments in pipe (due to internal pressure only). (Reproduced with the permission of the Controller of Her Majesty's Stationery Office, London).

The water contained nitrates that caused stress corrosion cracking of the mild steel reactor. Nitrate-induced cracking was well known to metallurgists but was not well known to other engineers at the time.[5]

Water has often been poured over equipment to cool it. Before doing so, however, we should ask what is in the water and what effect it will have on the equipment.

6. In a wider field, Lawless[6] examined 46 technological changes, from thalidomide to plastic turf, that had unforeseen and undesirable side effects. He concluded that in 40% of the cases the side effects could reasonably have been foreseen and that in 25% more notice might have been taken of early warning signs.

For other examples of modifications that went wrong and for accounts of ways of controlling modifications, see references 4 and 7–11.

Any modifications should be designed, constructed, tested, and maintained to the same standards as the original plant.

FROM THE OFFICIAL REPORT ON THE EXPLOSION AT FLIXBOROUGH ON JUNE 1st 1974

**FIGURE 25.5** One of the recommendations made to avoid another Flixborough.

69

*References*

1. Kletz, T.A. 1980. *Loss Prevention* 13:1.
2. *The Flixborough Cyclohexane Disaster.* 1975. London: Her Majesty's Stationery Office.
3. Lees, F.P. 1996. *Loss Prevention in the Process Industries,* 2nd ed., Appendix A1. Oxford, England: Butterworth-Heinemann.
4. Kletz, T.A. 1994. *Learning from Accidents,* 2nd ed., Chap. 8. Oxford, England: Butterworth-Heinemann.
5. Institution of Chemical Engineers. 1978. *Guide Notes on the Use of Stainless Steel in Chemical Process Plant,* p. 22. Rugby, England: Institution of Chemical Engineers.
6. Lawless, E.W. 1977. *Technology and Social Shock.* New Brunswick, NJ: Rutgers University Press.
7. Kletz, T.A. 1976. *Chem Engineering Progr.* 72(11):48.
8. Lees, F.P. 1996. *Loss Prevention in the Process Industries,* 2nd ed., Chap. 21. Oxford, England: Butterworth-Heinemann.
9. Institution of Chemical Engineers. 1994. *Modifications: The Management of Change.* Safety Training Package No. 025. Rugby, England: Institution of Chemical Engineers.
10. Kletz, T.A. 1994. *What Went Wrong—Case Histories of Process Plant Disasters,* 3rd ed., Chap. 2. Houston, TX: Gulf Publishing.
11. Sanders, R.E. 1993. *Management of Change in Chemical Plants—Learning from Case Histories.* Oxford, England: Butterworth-Heinemann.

# MYTH 26

## Plants are made safer by adding on protective equipment

The usual procedure in plant design is to

design the plant
identify the hazards
add on protective equipment to control the hazards

Typical protective equipment includes

gas detectors
emergency isolation valves
trips and alarms
relief valves and flare stacks
steam and water curtains

flame arrestors
nitrogen blanketing
fire protection equipment such as insulation and water sprays
fire-fighting equipment

Slowly, however, the industry is coming to realize that an alternative approach is possible: The best way of preventing a large leak of hazardous material is to use a safer material in its place, or to use less of the hazardous material, or to use it at lower temperatures and pressures (what you don't have can't leak). Such plants are inherently safer and do not have to be made safe by adding on protective equipment.

To use an analogy, if the meat of lions were good to eat, farmers would find ways of raising lions. Cages and other protective equipment would be required to keep them under control and only occasionally, as at Flixborough or Bhopal, would the lions break loose. But why keep lions when lambs will do instead?

Reference 1 describes some of the changes that have been or might be made, including the following.

1. Nitroglycerine used to be manufactured in a batch reactor containing 1 ton. Now it is made in a small continuous reactor containing about a kilogram, and the residence time has been reduced from 120 minutes to 2 minutes.
2. Adipic acid used to be made in a reactor fitted with external coolers. Now it is made in an internally cooled reactor, thus eliminating pump, cooler, and pipelines. Mixing is achieved by the gas produced as a by-product.
3. The new Higee distillation process reduces the quantity of material in the distillation equipment by a factor of up to 1000. Even in conventional distillation equipment, the inventory can be reduced by a factor of 2 or 3 by attention to detail.
4. By choice of suitable heat exchangers, the inventory can be reduced by a factor of 20 or more.
5. Water can be used as a heat-transfer medium instead of flammable oils. If flammable oils must be used, high-boiling-point oils are available (see Myth M2, Question 1).
6. The material that leaked at Bhopal in 1984, methyl isocyanate, which killed more than 2000 people, was not a product or raw material but an intermediate. While it was convenient to store

it, it was not essential to do so, and within a year after the accident the company involved, Union Carbide, and many other companies had drastically reduced their stocks of methyl isocyanate and other hazardous intermediates.

7. The leak at Flixborough (see Myth 25) was so large, and the explosion so devastating, because the conversion was low (about 6%) and the reactors contained a large amount of raw material, most of which (94%) got a "free ride" and had to be recovered and reprocessed.

   After the explosion, one company operating a similar plant tried to develop a more efficient process. The research was promising but was abandoned before completion, as there was no prospect that the company, or any other company in the West, would require a new plant in the foreseeable future.

Related to the concept of inherent safety is that of simpler plants. If we can spot hazards *early* in design, we may be able to *remove* them by a change in design instead of *controlling* them by adding on protective equipment.

An example is provided by Section 1.3 and Figure 1.1 of Myth 1. We may be able to avoid the need for relief valves and the associated flare system by using stronger vessels. However, the decision must be made early in the design process because vessels are normally ordered before the detailed design is complete. It is no use waiting until the hazard and operability study or relief and blowdown review is carried out late in the design process. By this stage, all we can do is to add on protective equipment, that is, relief valves, flare systems, etc.

*Reference*

1. Kletz, T.A. 1991. *Plant Design for Safety—A User-Friendly Approach.* Washington, DC: Taylor & Francis.

## MYTH 27

### The technical people know their jobs; the safety adviser can leave technical problems to them and concentrate on human failings

This myth was believed by many old-time safety advisers, partly because they were incapable of dealing with technical problems, but also because they genuinely believed that the technical staff had the technical problems under control. While this may have been true at one time, unfortunately, it is not true today, as is shown by some of the incidents described earlier, for example,

the fire at Feyzin (Myth 2)
fires and explosions involving high-boiling-point materials (Myth 9)
the vessel that was burst by compressed air because the vent was
    choked (Myth 16).
the modifications described under Myth 25.

Many low-pressure storage tanks have been sucked in because people did not realize how little vacuum the tanks could withstand. Tanks are usually designed to withstand a vacuum of only 2.5 inches of water gauge (0.1 psi or 0.6 kPa), equivalent to the hydrostatic pressure at the bottom of a cup of tea.[1] Tanks have been sucked in, for example, because flame arrestors were not cleaned, plastic bags were tied over vents to prevent dirt from getting in, or a loose blank was put over the vent to prevent fumes from coming out.

One of the jobs of the safety adviser in the process industries is to point out to his colleagues in design and production those facts that they ought to have learned early in their careers or, rather, to teach them to apply to practical situations the knowledge they have.

The safety advisor in the high-technology industries needs to be primarily[2]

a technologist, with knowledge and experience of the technology of the industry, able to hold his own with managers and designers.
a numerate person, able to apply systematic and numerical techniques.
a communicator in speech, writing, and discussions.
an auditor, able to spot incipient trouble by sharp eyes and regular surveys.

In addition, the safety advisor should understand the limitations of people and know when it is reasonable to rely on them and when to try to remove opportunities for error (see Myths 6–8).

*References*

1. Sanders, R.E. 1993. *Management of Change in Chemical Plants—Learning from Case Histories.* Oxford, England: Butterworth-Heinemann.
2. Kletz, T.A. 1982. *Safety and Loss Prevention.* Symposium Series No 73, p. B1. Rugby, England: Institution of Chemical Engineers.

## MYTH 28

## The public believes technology is making the world a better place

If people believed at one time that technology is making the world a better place, fewer seem to believe it today. Scientists and technologists are attacked by the media for producing both hazards and pollution. On the whole, we are not hitting back but waiting for the storm to blow over and wondering if the attacks are perhaps not partly justified. Many have lost the confidence in scientists that existed before and immediately after World War II, when it was taken for granted that science was the road to a better future.

This criticism of science and technology is not new. In *Victorian Engineering,*[1] L.T.C. Rolt wrote of the years following the Great Exhibition of 1851:

> The public began to lose confidence in the engineer so that he began to lose confidence in himself. The mood of uncritical reverence for material progress which had made the engineer the hero of the hour did not long outlive the Great Exhibition. The golden calf of the new Utopia, of man's emancipation by machine, was worshipped at the Crystal Place for the last time. A mood of doubt and disillusionment began, slowly at first, to undermine the old, easy confidence.

As an example of the new mood, Rolt said that the Forth Bridge was attacked as "the supremest specimen of all ugliness."

The scientist is now suffering the loss of popular esteem that the engineer experienced 100 years earlier.

Most of the opponents of science and technology are well meaning but often muddled or ill informed. It is sometimes difficult to communicate with them because of a lack of common ground and because, when illogicalities and inconsistencies are pointed out, we may be told that "A wise judgement contains no logic."[2]

Nevertheless, we should not ignore the comments of the opponents of technology, but point out the facts and hope, like Lord Rothschild, "that if some course of action is made self-evident by hard information, and if that course of action is not followed, you and I will say to the politician who denies it what Queen Elizabeth the First said to her man Cecil: 'Get out!'"[3]

Some of the facts that need to be pointed out are the following.

*conclude the advantages & disadvantages*

→ 1. *We cannot have the benefits of modern technology without some disadvantages in terms of pollution and safety.*

   It is true that some of the benefits are of doubtful value (striped toothpaste, for example), and the opponents of technology are quick to point this out. They are less ready to dispense with antiseptics and anesthetics.

2. *New technologies are usually less hazardous than old ones.*

   The Flixborough disaster has been described by a TV reporter at the time as "the price of nylon." Many must wonder if it was worth the risk, but we have to wear clothing of some sort, and the "accident content" of natural fibers such as wool and cotton is higher than that of man-made fibers. (The cost of an article is the cost of its labor content, capital costs being other people's labor. Natural fibers "contain" more labor because agriculture is a low-wage industry, and they "contain" more accidents because agriculture is a high-accident indus-
   → try.) Similarly, nuclear energy, despite Chernobyl, is safer than energy derived from oil or coal,[4-6] and plastics are safer to produce than metal or wood.

   One might have expected the environmental lobby to have favored nuclear energy because it does not produce acid rain, greenhouse gases, or large quantities of waste. Unfortunately,
   → nuclear power is associated with atomic bombs.

3. *The cost of reducing pollution and increasing safety has to be paid for in the end by the public.*

   There comes a time when even the most safety-conscious person objects to paying a lot of money for a further reduction to an already very small risk. The more we spend on safety, the less there is left to spend on reducing poverty and disease or on those goods and services that make life worth living, for ourselves and others.

→ 4. *People, not technology, create hazards and pollution.*

## T.C. Young wrote[7]

A human society has a technology, and technology does not exist outside of the context of a human society. To see pollution as the result of technology and nature out of balance is to shrink from reducing the equation to its lowest common denominator. To blame pollution on technology is the ultimate dodge of a society unwilling to take the blame for its own errors and stupidity. We are so accustomed to excusing everything from misbehaviour to crime, from social dislikes to murderous hates, on man's social environment rather than on man, that when we come to deal with the natural environment we refuse from habit to take the blame on ourselves. Instead we pretend that technology, our technology, is something of a life force, a will, and a thrust of its own, on which we can blame all, with which we can explain all, and in the end by means of which we can excuse ourselves.

Not so. It is people who make pollution, not technology.

The defenders of the environment are often defending the results of human's past activities in earning a living. The English countryside, for example, bears no resemblance to its appearance before the coming of humans and is as much a man-made feature as a factory.

Similarly, it is not computers and automation that cause unemployment but the way we use them.

*References*

1. Rolt, L.T.C. 1974. *Victorian Engineering,* pp. 162, 195. London: Penguin Books.
2. van Dieren, W. 1978. *Chem Industry* 2:899.
3. Lord Rothschild. *Risk,* p. 13. London: BBC Publications.
4. Health and Safety Commission. 1978. *The Hazards of Conventional Sources of Energy.* London: Her Majesty's Stationery Office.
5. Inhaber, H. 1978. *Risk of Energy Production.* Report No AECB-1119/REV-1. Ottawa, Canada: Atomic Energy Control Board.
6. Cohen, A.V., and Pritchard, D.K. 1980. *Comparative Risks of Electricity Production. A Critical Survey of the Literature.* Research Paper 11, Health and Safety Executive. London: Her Majesty's Stationery Office.
7. Young, T.C. 1975. In *Man in Nature,* ed. L.D. Levine, p. 10. Toronto, Canada: Royal Ontario Museum.

# MYTH 29

## Major failures of plants and equipment are now so infrequent that it is rarely possible to reduce them further

People spend a great deal of money and effort estimating the *probability* of major failures and the *consequences* of these failures, but very little effort considering their *causes* and methods of prevention.

For example, in 1982 the Institution of Chemical Engineers organized a symposium on The Assessment of Major Hazards.[1] The 24 papers presented divided about equally into those that discussed the *probability* of a major incident and those that discussed the *consequences*. Not one paper discussed the causes. It was as if the occasional major leak was considered inevitable.

At the Fourth International Symposium on Loss Prevention[2] held in England in 1983, more than 100 papers were presented. Over half considered methods of estimating the probabilities and consequences of major leaks. About a dozen papers described incidents that have occurred and made specific recommendations to prevent them from happening again, but there were no general papers on the causes of leaks.

At the Seventh International Symposium on Loss Prevention held in Italy in 1992, 178 papers were presented; only 7 described incidents that have occurred, and again, there were no general papers on the causes of leaks (though there were a number on the underlying weaknesses in management that lead to accidents).[3]

Expensive experiments have been carried out to determine the behavior of major leaks—how they disperse in the atmosphere and how they behave when ignited.[4] In contrast, little effort has been devoted to studying the causes of leaks, though doing so would cost a fraction of the amount spent on studying the behavior of leaks. Yet if we could prevent leaks, we would not have to worry so much about the behavior of the leaking material.

Why has there been this imbalance in the allocation of resources? Do we prefer to tackle the more interesting or more fashionable problems—instead of the most important ones? This is often looked upon as an academic failing, but most of the experiments I have described have been carried out by industry— or paid for by them.

My own examination of industry reports suggests that the biggest single cause of major leaks[5-7] is pipe failure and that many of these failures arise at the design-construction interface—the construction team does not follow the design, or the details are left to them and are not carried out in accordance with good engineering practice. To reduce the number of major leaks, we should specify designs in detail and then check after construction to make sure that the design has been followed and that details not specified have been carried out in accordance with good engineering practice.

Here are some examples of accidents caused by a failure to follow designs in detail or to do well what has been left to the discretion of the construction team:

1. Small-bore branches have been inadequately supported, and have vibrated and failed by fatigue.
2. Old pipe that was already corroded or had used up some of its creep life has been reused (see Myth 47).
3. Temporary supports have been left in position, or temporary branches, installed for pressure testing, have not been welded up.
4. The end of a relief valve tailpipe has been placed so close to the ground that it was sealed by a frozen puddle.
5. Dead-end branches have been left in pipe work, and have filled with water (which froze, breaking the line) or with corrosive by-products.
6. Bellows have been installed between pipes that were not in line, so that the bellows were distorted.

*References*

1. Institution of Chemical Engineers. 1982. *The Assessment of Major Hazards.* Symposium Series No 71. Rugby, England: Institution of Chemical Engineers.
2. Institution of Chemical Engineers. 1983. *Loss Prevention and Safety Promotion in the Process Industries.* Symposium Series Nos 80–82. Rugby, England: Institution of Chemical Engineers.
3. SRP Partners. 1992. *Seventh International Symposium on Loss Prevention in the Process Industries.* Rome, Italy: SRP Partners.
4. Pikaar, M.J. 1983. In *Loss Prevention and Safety Promotion in the Process Industries.* Symposium Series Nos 80–82, Vol. 1, p. C20. Rugby, England: Institution of Chemical Engineers.
5. Kletz, T.A. 1980. Safety aspects of pressurized systems. In *Proceedings of the Fourth International Conference on Pressure Vessel Technology,* p. 25. London: Institution of Mechanical Engineers.
6. Kletz, T.A. 1983. *Plant/Operations Progr* 3(1):19.
7. Kletz, T.A. 1994. *Learning from Accidents,* 2nd ed., Chap. 16. Oxford, England: Butterworth-Heinemann.

## MYTH 30

### If equipment has to be opened frequently, then quick-release couplings should be fitted to save time and effort

Every day, in every process factory, equipment that has been under pressure is opened for cleaning or repair. Normally, a process operator prepares the equipment by blowing off the pressure and, if necessary, removing any traces of the contents that remain. The operating department then issues a work permit to a maintenance worker, who opens the equipment, usually by unbolting the bolts that hold the cover or flanges in place. If the bolts are slackened correctly, any pressure left in the equipment is immediately apparent, and the bolts can be tightened or the pressure allowed to blow off.

Safety is achieved by (1) a procedure that makes one person responsible for preparation of the job and another responsible for actually carrying it out, and (2) an opening technique that can cope with a failure to carry out the preparation correctly.

When equipment has to be opened frequently, it is sometimes designed so that the entire operation can be carried out by the operator, using quick-release couplings. Sooner or later, by oversight or neglect, an attempt is made to open the equipment before the pressure has been blown off. Several accidents that have occurred in this way are described below.

If quick-release couplings are used, the following procedures are necessary.

1. Interlocks should be provided, so that the equipment cannot be opened until the source of pressure is isolated and the vent valve opened.
2. The design of the door or cover should allow it to be opened about ¼ inch (6 mm) while still capable of carrying the full pressure, and a separate operation should be required to release the cover fully. When the cover is cracked open, any pressure in the vessel is immediately apparent, and the cover can be resealed or the pressure allowed to blow off through the gap.

Experience shows that Item 1 alone is not sufficient, especially if the vent is liable to choke.

*Accidents Resulting from the Opening of Pressure Vessels Using Quick-Release Couplings*

1. During the 1960s, a suspended catalyst was removed from a process stream in a pressure filter. After filtration was complete, steam was used to blow the remaining liquid out of the filter. The pressure in the filter was blown off through a vent valve, and the fall in pressure was observed on a pressure gauge. The operator then opened the filter for cleaning. The filter door was held closed by eight radial bars that fitted into U-bolts on the filter body. The bars were withdrawn from the U-bolts by turning a large wheel, fixed to the door. The door could then be withdrawn.

   One day an operator started to open the door before blowing off the pressure. He was standing in front of it and was crushed between the door and part of the structure, and was killed instantly. In this sort of situation, sooner or later, through oversight or neglect, an attempt will be made to open the equipment while it is under pressure; on this occasion the operator was at the end of his last shift before going on holiday. It is too simple to say that the accident was due to the operator's error. The accident was the result of a work situation that made an accident almost inevitable. However, this was not fully recognized at the time, and less change was made to the plant than we would make today[1] (see also Myth 16).

2. Plastic pallets were being blown out of a tank truck by compressed air. When the tank seemed empty, the driver opened a manhole cover on the top to make sure. One day a driver, not a regular, started to open the manhole before releasing the pressure. When he had opened two of the quick-release couplings, the cover was blown open. The driver was blown off the tank and killed by the fall.

   Either the driver forgot to vent the tanker or he thought that it would be safe to let the pressure (10 psig) blow off through the manhole. After the accident, the manhole covers were replaced by the type described above, and the vent valve was moved from the side of the tanker to the foot of the ladder.[1]

   Many of those concerned were surprised that 10 psig could cause so much injury (see Myth 16).

**FIGURE 30.1** Vacuum flask stopper. Any trapped pressure will be released before the stopper is unscrewed.

3. An operator was using a portable sprayer that was pressurized by a plunger-type pump. The spray reduced to a dribble, and the sprayer seemed light in weight, so he opened it to refill it. To do so, he had to tap the screwed top loose with a hammer. The top came away with sufficient force to knock him off his feet. He had not used this type of sprayer before.

Investigation showed that the reduced spray was due to a plugged nozzle. The sprayer was scrapped and replaced by one fitted with a two-stage opening device, as described above, and a pressure gauge.

4. On a new plant a large filter on a 14-inch (35-cm) liquid propane line was fitted with a process-operated filter. The filter was opened while still under pressure, and two men were hospitalized with cold burns.

The company concerned had agreed some years before that all pressure vessels that can be opened by operators should be fitted with two-stage opening devices. However, filters were classified as "pipe fittings" rather than vessels and were designed by the piping department, which was not familiar with the vessel specifications.

With small filters (containing, say, only a few kilograms of liquid) the same precautions necessary with vessels may not be required, but they are necessary with large filters.

*An Illustration from Another Branch of Technology*

Figure 30.1 shows the stopper from a vacuum flask. It is hollow, and there is an opening in the wall of the threaded section. The stopper has been made this way to reduce heat transfer, but it has another advantage. When the stopper is unscrewed a few turns, any trapped pressure is immediately apparent. It can be allowed to blow off or the stopper screwed back. The contents of the flask cannot be ejected.

*Reference*

1. Kletz, T.A. 1991. *An Engineer's View of Human Error*, 2nd ed. Rugby, England: Institution of Chemical Engineers.

## MYTH 31

### The transport of chemicals is inevitably dangerous

At San Carlos de la Rapita in Spain in 1978 more than 200 people were killed when a tank truck of propylene disintegrated and the contents ignited.[1,2] In Germany, in 1943, 209 people were killed when a tank car of ether burst and the contents exploded.[3] In New York State in 1958, 200 people were injured when a tank car of nitromethane exploded.[4] There have, however, been no incidents of this magnitude in recent years.

In the United Kingdom the road transport of chemicals and gasoline killed 16 people during the period 1970–82, an average of 1.2 per year[5,6]; half of those killed were drivers, and half members of the public; half were killed by chemicals, and half by gasoline. Since 1982 the number of deaths per year has decreased. This statistic excludes accidents in which people were killed by mechanical impact and not by the contents. (Tank trucks, it is worth noting, are involved in fewer such accidents than other heavy vehicles.)

Forty-six people were killed in the United States by accidents involving tank cars (excluding mechanical impact) during the period 1969–78, compared with none in the United Kingdom[7]; since 1978, far fewer people have been killed in the United States (see item 3 below), and none in the United Kingdom.

Has the United Kingdom been lucky, or is its good record due to higher standards? At least some of the difference is probably due to higher standards, for the following reasons.

1. In Europe, steel used for some tank trucks is of types that would not be used (for certain loads) in the United Kingdom. The tank involved in the Spanish disaster was made from T1 steel and had been used for carrying ammonia. As a result, stress corrosion cracking had occurred. A similar tank failed in France in 1968 while being filled with ammonia, and five people were killed.[8,9]
2. In the United Kingdom, tank trucks carrying liquefied flammable gases under pressure are fitted with relief valves, but in Europe, this is not the case. Relief valves would probably have prevented the German incident and many lesser incidents.
3. In the United States the railway tracks are in much poorer condition than in the United Kingdom, and derailments are more frequent. Derailments have caused many tank accidents. Other accidents have been due to the free shunting of tank cars, a practice not permitted in the United Kingdom. U.S. tank cars have now been fitted with improved couplers, with head shields to prevent couplers piercing the end of the tank cars, and with insulation to protect the tank cars from fire. These measures have reduced the numbers of people killed.[7] No one

seems to have asked if it would have been cheaper to improve the track.

To sum up, there is some truth to the myth, in that the transport of chemicals has at times been dangerous in some countries; however, it does not have to be so.

### References

1. Stinton, H.G. 1983. *J Hazardous Materials* 7:373. (There are a number of inaccuracies in this account.)
2. Hymes, I. 1983. *The Physiological and Pathological Effects of Thermal Radiation.* Report No SRD R 275, Appendix 5. Warrington, England: UK Atomic Energy Authority.
3. Lewis, D.J. 1983. *Hazardous Cargo Bull* 4(8):12.
4. Lewis, D.J. 1983. *Hazardous Cargo Bull* 4(9):36.
5. Kletz, T.A. 1984. *Hazardous Cargo Bull* 5(11):10.
6. Kletz, T.A. 1986. *Plant/Operations Progr* 5(3):160.
7. US National Transportation Safety Board. 1979. *Safety Report on the Progress of Safety Modifications of Railroad Tank Cars Carrying Hazardous Materials.* Washington, DC: U.S. Government Printing Office.
8. *Annales des Mines,* January 1969, p. 25.
9. *Ammonia Plant Safety* 1970. 12:12.

## MYTH 32

### Major disasters in the chemical industry are becoming more frequent

The impression we get from the media is that chemical disasters are becoming more frequent, and it is supported by such statements as the following:

> The trends in frequency and proportion of incidents producing blast and fatalities in UVCE's (unconfined vapour cloud explosions) . . . are all upwards.[1]

> . . . the number of incidents is on the increase despite the activities of the past few years.[2]

> . . . there is little evidence that the oil crisis has checked the increase in rate of occurrence of serious incidents.[3]

> The impression was that the frequency of disasters on an international scale was increasing.[4]

On the other hand, another author wrote

> By almost every type of societal indicator, except one, hazardous events have
> been increasing. . . . The one exception is the statistical record of hazard
> consequences . . . there frequently appears an enormous divergence between
> this record and the perception of hazards by scientists, the public and
> officialdom.[5]

During the 1970s, I kept records of serious incidents in the oil
and chemical industries that were reported in the press and in sources
such as references 1 and 6. From these, I extracted details of all
incidents in which five or more people were killed. They are summarized in Table 32.1, which shows that the number of incidents
occurring on fixed installations (plants, storage areas, pipelines) was
roughly constant during the period covered (1970–80) despite the
growth in the size of the industry, which more than doubled during
the decade. For fixed installations, the average number of incidents

**TABLE 32.1**  Annual numbers of serious incidents and deaths in the oil and chemical
industries

| Year | Total | | Fixed installations | | Transport | |
|---|---|---|---|---|---|---|
| | No. of incidents | No. killed | No. of incidents | No. killed | No. of incidents | No. killed |
| 1970 | 7 | 127 | 5 | 115 | 2 | 12 |
| 1971 | 6 | 545 | 4 | 505 | 2 | 40 |
| 1972 | 12 | 364 | 6 | 136 | 6 | 228 |
| 1973 | 10 | 140 | 8 | 114 | 2 | 26 |
| 1974 | 7 | 189 | 5 | 166 | 2 | 23 |
| 1975 | 10 | 114 | 8 | 68 | 2 | 46 |
| 1976 | 15 | 156 | 13 | 141 | 2 | 15 |
| 1977 | 11 | 123 | 6 | 55 | 5 | 68 |
| 1978 | 17 | 483 | 8 | 99 | 9 | 384 |
| 1979 | 17 | 324 | 8 | 147 | 9 | 177 |
| 1980 | 9 | 137 | 6 | 71 | 3 | 66 |
| TOTAL | | | | | | |
| (11 years) | 121 | 2702 | 77 | 1617 | 44 | 1085 |
| Mean | | | | | | |
| (per year) | 11 | 246 | 7 | 147 | 4 | 97 |
| Fatalities/ | | | | | | |
| Incident | — | 22 | — | 21 | — | 24 |

was 7 per year with a range of 4-13. The average number of people killed was 147 per year with a range of 55-505.

Transport accidents are a different matter. For the period 1970-77, the average number of incidents was 3 per year with a range of 2-6. The average number of people killed was 57 per year with a range of 12-228. However, in 1978 there were 9 incidents and 384 deaths, and in 1979 there were 9 incidents and 177 deaths. The year 1978 was dominated, of course, by the Spanish tank truck disaster (see Myth 31), but even without this incident, it would have been a black year for the transport of oils and chemicals, exceeded only by 1972. Only one of the transport incidents occurred in the United Kingdom (in 1979).

My collection of the press accounts was not exhaustive but was more thorough in the later years and therefore would tend to bias any growth in the number of incidents.

My definition of the oil and chemical industries was very broad. It included accidents caused by conventional explosives, an incident in which people were killed while collecting gasoline from a leaking tank truck, an incident in which people were poisoned by eating stolen seed corn that had been treated with a poisonous insecticide, and explosions due to leaks from natural gas transmission lines. Details are given in reference 7.

Keller and Wilson[8] reviewed similar data for the period 1980-87 and found a small decrease in the incident rate, compared with that of 1970-79. However, the change is not significant, as data collection may not have been as thorough as in the earlier period. The actual number of deaths per year during the 1980s was certainly higher than in the 1970s, as more people were killed by three 1984 incidents (the release of toxic gas at Bhopal, India; the LPG fire at Mexico City, and a pipe failure and fire at Cubatao, near Sao Paulo, Brazil) than by all incidents combined in the 1970s.

To summarize, during the 1970s the chemical industry doubled in size, but the number of serious incidents and the number of deaths per year did not rise. In the 1980s the number of incidents seems to have continued at much the same rate, but there were three very serious incidents accounting, between them, for 3000 deaths (more, according to unofficial estimates).

*References*

1. Gugan, K. 1978. *Unconfined Vapour Cloud Explosions,* p. 103. Rugby, England: Institution of Chemical Engineers.

2. *Processing,* April 1979, p. 25.

3. Marshall, V.C. 1979. Quoted in *Processing,* April, p. 25.

4. Barrell, A.C. 1979. Quoted in *The Guardian,* 24 April.

5. Kates, R.W. 1978. *Managing Technological Hazard—Research Needs and Opportunities.* Boulder, CO: University of Colorado.

6. Nash, J.R. 1977. *Darkest Hours.* New York: Pocket Books.

7. Turner, E., and Kletz, T.A. 1979. *Is the Number of Serious Accidents in the Oil and Chemical Industries Increasing?* London: Chemical Industries Association.

8. Keller, A.Z., and Wilson, H.C. 1991. *Hazards XI—Directions in Process Safety.* Symposium Series No 124, Rugby, England: Institution of Chemical Engineers.

# MYTH 33

## Current methods must be safe because we have done it this way for years without an accident

Experience, "practical people" say, shows that an operation has been carried out in a particular way for 20 years without an accident. That proves it is safe.

What do we mean by safe? Is an accident in the 21st year acceptable? If it is not, then we have not proved that the operation is safe because an accident might occur at any time.

Even if one accident in 20 years is acceptable, we have not even shown that on *average* the accident rate is less than one in 20 years. Suppose that *on average* one accident occurs in a period of time such as 20 years; then on average there is a 37% chance that no accidents will occur in this period of time. Suppose on average two accidents occur in a period of time; then on average there is a 14% chance that no accidents will occur in this period of time.

These figures are derived from the Poisson distribution, which shows that if the average number of random events that will occur in a given time is $\mu$, then the probability $P$ that $n$ events will occur is given by

$$
\begin{array}{ccccc}
\text{No. of events} = & 0 & 1 & 2 & n \\
\text{Probability} = & e^{-\mu} & e^{-\mu}\mu & e^{-\mu}\dfrac{\mu^2}{2!} & e^{-\mu}\dfrac{\mu^n}{n!}
\end{array}
$$

If $\mu = 1$ and $n = 0$, then $P = e^{-1} = 0.37$.
If $\mu = 2$ and $n = 0$, then $P = e^{-2} = 0.14$.

Putting it the other way around, if no failure has occurred in 20 years, what is the probability that the average failure rate is 1 in 10 years or less?

If the average failure rate is 1 in 10 years, then the probability of no failure in 10 years is 0.37, and the probability of no failure in the following 10-year period is also 0.37. The probability of no failure in a 20-year period is $0.37 \times 0.37 = 0.14$. If there has been no failure in a 20-year period, we can be 86% confident that the average failure rate is 1 in 10 years or less.

Let us look at some accidents that occurred because, on the basis of past experience, people assumed that a plant was safe. (See also Myth 13.)

1. At the BASF plant at Oppau, Germany, in 1921, explosives were used to break up storage piles of a 50:50 mixture of ammonium sulfate and ammonium nitrate. Two terrific explosions occurred, killing 430 people, including 50 members of the public, destroying the plant and 700 houses, and producing a crater 250 feet (75 m) across and 50 feet (15 m) deep. The operation had been carried out without mishap 16,000 times before the explosion occurred.[1]
2. While starting up a coker—a plant for making coke—an operator forgot to open a valve. An explosion occurred, and a man was killed. The plant and another similar one were each started up every few days because they had to be shut down for emptying between batches, and there had been 6000 successful start-ups before the explosion occurred.[2]
3. Nitromethane was considered nonexplosive and safe to transport in tank cars from 1940 until 1958, when two tanks exploded in separate incidents in the United States. Both occurred as the result of shunting. Although only two people were killed, many were injured, damage was extensive, and the second incident produced a crater 100 feet (30 m) across and 33 feet (10 m) deep.[3]
4. H.A.L. Fisher, describing the fall of Constantinople in 1453, wrote, ". . . since the city had never been taken, a belief prevailed that it could not be taken. Cities, however, are not defended by beliefs but by material power."[4]

Why do accidents occur after a long period of safe operation? In some cases, conditions were never exactly right for an accident. A blind man may walk along the edge of a precipice for a long time before he falls off. A software error in a computer-controlled plant may lie like a time bomb until a particular fault develops at a particular stage of a batch.[5] In other cases, a gradual change in operating conditions or scale, not recognized as significant (see Myth 25), may produce a hazard. For example, oil fields that produce sweet products (that is, oil and gas free from hydrogen sulfide) can gradually become sour. If this is not detected in time and action taken, there can be unexpected corrosion.

*References*

1. Lees, F.P. 1996. *Loss Prevention in the Process Industries,* 2nd ed., Appendix 3. London: Butterworth-Heinemann.
2. Vervalin, C.H. 1985. *Fire Protection Manual for Hydrocarbon Processing Plants,* 3rd ed., p. 95. Houston, TX: Gulf Publishing.
3. Lewis, D.J. 1983. *Hazardous Cargo Bull* 4(9):36.
4. Fisher, H.A.L. 1936. *A History of Europe,* p. 408. London: Edward Arnold.
5. Kletz, T.A., et al. 1995. *Computer Control and Human Error.* Rugby, England: Institution of Chemical Engineers; and Houston, TX: Gulf Publishing.

## MYTH 34

### If there were five accidents last year and seven the year before, then the accident rate is getting better

In industry we often see reports in which a factory is praised for a small decrease in the numbers of accidents or blamed for a small increase. Is the change real, or is it just due to chance? Figures 34.1 and 34.2 may help us decide.[1,2] Two cases are considered.

*The Average (or Expected) Number of Incidents Is Known*

Suppose that on average there are nine accidents (or nine leaks or nine fires) per month (or per week or per year). Suppose that in a particular month (or week or year) there are only four. Has the accident record really improved?

Looking at the value of 9 on the horizontal axis of Figure 34.1 and then following the vertical line upward, we see that it crosses the 90% confidence line at the value of 4. This means that we can be 90% certain that there has been an improvement. If we say things have improved, 9 times out of 10 we will be right.

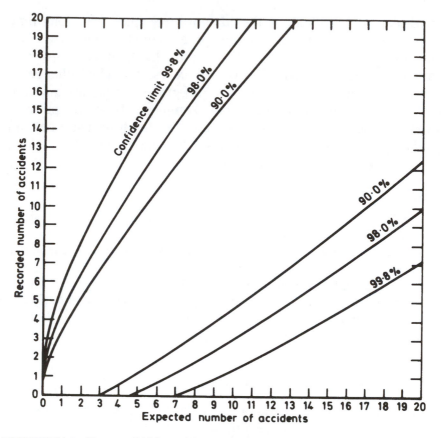

**FIGURE 34.1**   How to tell if the accident record has really improved.

If, however, there are five incidents one month, we ought to wait another month or two before congratulating anyone on an improvement.

If instead of four incidents there are only two, we can be 98% certain that things have improved, that is, 98 times out of 100 we will be right.

If there is only one incident, we can be 99.8% certain that things have gotten better; 998 times out of 1000 we will be right.

Similarly, if there are 15 incidents, we can be 90% certain that things are worse. If there are 17 incidents, we can be 98% certain, and if there are 20 incidents, we can be 99.8% certain.

*The Average Number of Incidents Is Not Known*

So far we have assumed that we know the average (or expected) incident rate. Suppose there were 20 incidents in one period and 10

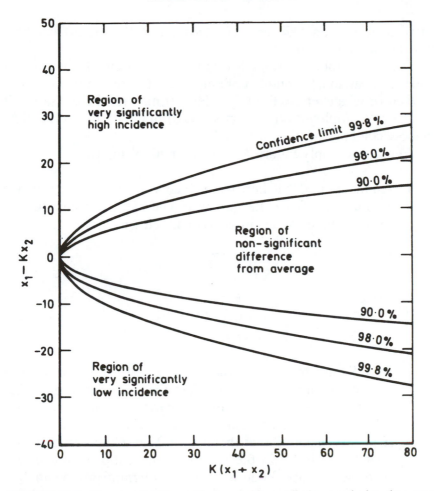

**FIGURE 34.2** How to tell if the accident record has really improved when the average record is not known.

in the next and we do not know the average. Figure 34.2 can then be used. If

$$x_1 = 20$$
$$x_2 = 10$$
$$x_1 + x_2 = 30$$
$$x_1 - x_2 = 10$$

find $x_1 + x_2$ on the bottom scale of Figure 34.2 and move upward. We cross the 90% line at $x_1 - x_2 = 9$. Here, $x_1 - x_2$ is actually equal to 10, so we are more than 90% certain (but less than 98% certain) that there has been an improvement.

In this example, $K = 1$. $K$ allows for the fact that in the two periods under comparison, the number of employees (or the number of hours worked or whatever we think is relevant) may not be the same. If there are $N_1$ employees (or hours worked) in the first period and $N_2$ in the second period, then $K = N_1/N_2$.

Let us now apply Figure 34.2 to two real examples.

1.  A unit reported that there were seven lost-time accidents in one year but only five in the next year. The staff was congratulated on a "commendable" improvement. However, if

$$x_1 = 7$$
$$x_2 = 5$$
$$K = 1$$
$$x_1 + x_2 = 12$$
$$x_1 - x_2 = 2$$

from Figure 34.2 we see that if $x_1 + x_2 = 12$, $x_1 - x_2$ must be 5 before we can be 90% confident that an improvement has occurred. There is thus no evidence for a real improvement; a fall from seven to five could be due to chance.

2.  A factory reported that there were 93 road accidents in one year but only 86 in the following year, though traffic had increased by 5%. Road users were congratulated on an "encouraging" improvement. However, if

$$x_1 = 93$$
$$x_2 = 86$$
$$K = 0.95$$
$$K(x_1 - x_2) = 0.95 \times 179 = 170$$
$$x_1 - Kx_2 = 93 - 82 = 11$$

On Figure 34.2, the value of 170 is off the scale, but using a similar graph, we find that $(x_1 - Kx_2)$ must be 20 or less if we are to be 90% confident there has been an improvement. There is thus no evidence that the road accident record has really improved—the change may be due to chance.

*Notes*

1. The graphs cannot be used for comparing accident *rates;* they can only be used for comparing numbers of accidents (or other events).
2. Any conclusions drawn from Figure 34.2 will be wrong if the wrong value of $K$ is used. For example, we might assume that the number of accidents depends on the number of hours worked and take $K$ as the ratio between the number of hours in the first period and the number in the second. If the number of accidents is actually proportional to output, our conclusions may be wrong.

*References*

1. Wynn, A.H.A. 1950. *Applications of Probability Theory to the Study of Mining Accidents.* Report No 7. Sheffield, England: Safety in Mines Research Establishment.
2. Lees, F.P. 1996. *Loss Prevention in the Process Industries,* 2nd ed., Chap. 27. London: Butterworth-Heinemann.

*Appendix to Myth 34: The Manager's Dream*

The following story shows how easily we can be misled by accident statistics—when the numbers involved are small.

It was an evening at the end of January 1994 and the Division Manager could put it off no longer. He had to get out his safety report for the previous year. The report would show that his division had had six lost-time accidents—the same as in the year before. Every other division in the group had shown an improvement—between 10% and 40%—but his record had remained the same. The Board had said that in 1993 the lost-time accident record was one of the things on which Division Managers would be assessed. It was rumored that the Division Manager showing the biggest improvement would be promoted, the one showing the worst. . . .

He gazed at the figures showing accidents month by month, hoping that somehow one of them would disappear.

|       | 1982 | 1993 | 1994 |
|-------|------|------|------|
| Jan   | 0    | 2    | 0    |
| Feb   | 0    | 0    |      |
| March | 1    | 0    |      |
| April | 0    | 1    |      |
| May   | 1    | 0    |      |

| | | |
|---|---|---|
| June | 0 | 1 |
| July | 1 | 0 |
| Aug | 0 | 1 |
| Sept | 0 | 1 |
| Oct | 2 | 0 |
| Nov | 0 | 0 |
| Dec | 1 | 0 |
| | | |
| Total | 6 | 6 |

While the Division Manager was working, his wife was watching television. He looked up to see the President announce that in the future the year would start on February 1 instead of January 1.

Looking at his figures, the Division Manager saw that his accident record showed a remarkable improvement—from 8 in 1992 to 4 in 1993—he had halved the number of accidents.

| | 1992 | 1993 | 1994 |
|---|---|---|---|
| Jan | 0 | 2 | 0 |
| Feb | 0 | 0 | |
| March | 1 | 0 | |
| April | 0 | 1 | |
| May | 1 | 0 | |
| June | 0 | 1 | |
| July | 1 | 0 | |
| Aug | 0 | 1 | |
| Sept | 0 | 1 | |
| Oct | 2 | 0 | |
| Nov | 0 | 0 | |
| Dec | 1 | 0 | |
| | | | |
| Total | ~~6~~ | ~~6~~ | |
| | 8 | 4 | |

He always knew he could improve the safety record if he set his mind to it.

"Wake up dear," said his wife. "You've fallen asleep again."

All that the Division Manager had to do was find a convincing reason for starting the safety year in February rather than January. That should not be too difficult—the Safety Committee took office on February 1; the safety competitions started on February 1. They had not done so in the past, but that could soon be changed. As for next year—he looked forward to a promotion, and his successor could worry.

# MYTH 35

## The effectiveness of a relief valve is not affected by the position of the vessel

Plant vessels do not usually move, but tankers may topple over as the result of an accident. The relief valve may then be below the liquid level, so that it discharges liquid, not vapor. If the tanker contents are flammable, the relief valve will have been sized to pass the volume of vapor produced when the tanker is surrounded by fire. To prevent the tanker from being overpressurized and possibly bursting, the relief valve would have to pass a volume of liquid equal to the volume of vapor produced, and it is, of course, much too small to do this.

This cause of bursting is quite different from that which occurs in a boiling liquid expanding vapor explosion (BLEVE) (see Myth 2), when the metal above the liquid level is softened by fire and loses its strength.

# MYTH 36

## An oversized relief valve will provide increased safety

If the valve is far too big, it may chatter, and the vibration may cause a joint to leak. Two fires that started in this way are described in references 1 and 2. The vibration was particularly severe, as there were two relief valves in one case and three in the other, and they competed with each other and cycled rapidly. If two or more relief valves have to be installed, they should be set at slightly different pressures.

In the first incident, two 6-inch (15-cm) relief valves lifted because there was a delay, after start-up, in opening a downstream isolation valve. Within 30 seconds, a 6-inch (15-cm) flange started to leak, and the escaping oil ignited.

In the second incident the relief valves lifted when the wrong isolation valve was operated. Vibration loosened the bolts around the flanged connection and cracked a welded joint. The leaking gas ignited, destroying the surrounding equipment.

*References*

1. Politz, F.C. 1985. Poor relief valve piping design results in crude unit fire. Paper presented at American Petroleum Institute 50th Mid-year Refining Meeting, Kansas City, MO.

2. Hancock, B. 1989. In *Safety and Loss Prevention in the Chemical and Oil Processing Industries,* p. 589. Symposium Series No 120. Rugby, England: Institution of Chemical Engineers.

## MYTH 37

### A few inches of water in the bottom of a heavy fuel oil tank will not cause any trouble

If oil above 100°C is added to the tank, the water will vaporize with explosive violence, the vent will probably not be big enough to pass the mixture of steam and oil, and the roof will be blown off the tank.

Alternatively, if the oil in the tank is heated above 100°C, the heat will gradually travel through the oil to the water layer. When the water boils, the steam will lift the oil, reducing the pressure, and the water will boil with increased vigor. Again the mixture of steam and oil will probably blow the roof off the tank. In one incident a structure 25 m (80 ft) tall was covered with oil. Witnesses said that the tank had exploded, but it was a physical explosion they had seen rather than a chemical one.

The phenomenon is known as foam-over, froth-over, boil-over, slop-over, or puking. The term boil-over is usually used if the tank is on fire and hot residues from the burning layer travel down to the water layer. The term slop-over is often used if water from fire hoses vaporizes as it enters a burning tank.[1]

To prevent foam-overs, which are very common in the oil industry, the incoming oil must be kept below 100°C. It is good practice to fit a high-temperature alarm on the oil inlet line.

Alternatively, the tank contents must be kept well above 100°C, so that any small quantities of water that enter are vaporized and a water layer never builds up. In addition, the tank should be drained regularly and the contents circulated before any fresh oil is added. Addition of fresh oil should start at a low rate, and not more than a tenth of the tank capacity should be added at a time. The temperature of a heavy oil tank should never be allowed to fluctuate above or below 100°C.

Foam-over has been used to illustrate a possible extension of computer control. If the temperature of the incoming oil or the existing oil in a tank approaches 100°C, then the screen could display a warning message, not merely announcing a high temperature but reminding the operator of the consequences. The reminder message could also be displayed if the operator starts up or increases the heat supply to a tank that contains a water layer. A "help" function could explain why the consequences may occur and refer the operator to a plant instruction, accident report, or other document for more information. These documents could be accessible on the screen.[2,3]

*References*

1. *Loss Prevention Bulletin,* No 057, June 1984, p. 26; No 059, Oct 1984, p. 35; and No 061, Feb 1985, p. 36.
2. Newton, A.M. 1995. *Loss Prevention Process Industries* 8(1):41.
3. Kletz, T.A., Broomfield, E., Chung, P.W.H., and Shen-Orr, C. 1995. *Computer Control and Human Error,* p. 39. Rugby, England: Institution of Chemical Engineers; and Houston, TX: Gulf Publishing.

*Further Reading*

Kletz, T.A. 1994. *What Went Wrong—Case Histories of Process Plant Disasters,* 3rd ed., Sections 9.1.1, 12.2, and 12.4.5. Houston, TX: Gulf Publishing.

## MYTH 38

### If you can see the bottom of a hole in the ground, it is safe to go into it

Many workers have been overcome and killed because they entered pits, depressions, or other confined spaces without adequate testing. A tombstone in St. Cuthbert's Church, Marton, Middlesbrough, UK, reads

> Erected in memory of Robert Armstrong aged 28, James Ingledew aged 39 and Joseph Fenison aged 27 years who unfortunately lost their lives on Oct 11th 1812 by venturing into a well at Marton when it was filled with carbonic acid gas or fixed air. From this unhappy incident let others take warning not to venture into wells without first trying whether a candle will burn in them; if the candle burns to the bottom they may enter with safety; if it goes out human life cannot be supported.

According to the parish magazine, the three men entered the well to recover some stolen beef they had hidden there.

A more recent incident[1] is typical of many. Some diesel fuel tanks were located in a pit. A man went down a ladder to drain water from the tanks. The job was done regularly, but there were no instructions, no tests, and no breathing apparatus. The man collapsed. Another man entered to rescue him and also collapsed. The first man recovered, but the second one died.

Before anyone enters a vessel or other confined space that may contain dangerous vapor or fumes, the atmosphere should be tested and the confined space isolated from all sources of danger. If necessary, air masks should be worn. Entry should be authorized in writing by a responsible person. In many countries these precautions are required by law.

In 1984 there was an explosion in a water pumping station at Abbeystead, in which 16 people were killed. No one knew that methane might be present, and so no precautions were taken.[2] However, it is surprising that people were allowed to enter an underground pumphouse without first testing the atmosphere for carbon dioxide.

*References*

1. *Health and Safety at Work* 1984. April, p. 10.
2. Health and Safety Executive. 1985. *The Abbeystead Explosion*. London: Her Majesty's Stationery Office.

## MYTH 39

### Fire is worse than smoke

Fire causes more damage than smoke; flash fires and BLEVEs (see Myth 2) kill people before the smoke has had time to affect them. However, when fires occur in buildings, aircraft, and trains, more people are killed by carbon monoxide and other toxic chemicals in the smoke than by the fire itself. Several companies are now producing smoke hoods, which are designed to let people leave smoke-filled buildings and aircraft in safety. However, some of those on offer do not provide protection against carbon monoxide. At the time of this writing, Germany is the only country, so far as I am aware, with a standard and an approval system for the hoods.

## MYTH 40

### Endothermic reactions are safe because they cannot run away

In endothermic reactions the absorption of energy tends to reduce the temperature, and heat usually has to be supplied to the reactor to keep the reaction going. Therefore it is correct to say that endothermic reactions cannot run away. However, this does not mean that they are safe, as the products of endothermic reactions have a high bond energy, from energy put into them, and tend to be unstable. Typical endothermic compounds are acetylene, hydrogen cyanide, hydrazine, nitrogen trichloride, and butadiene. Bretherick[1] gives an extensive list. Benzene and toluene are endothermic but stable, owing to the effects of resonance.

In addition, with endothermic reactions, as with exothermic reactions, a deliberate or unwitting change in reaction conditions can result in a different reaction taking place, and this reaction may be violent. For example, for many years a reaction had run smoothly in the presence of powdered sodium carbonate (soda ash), which neutralized an acidic by-product. "Natural soda" was used instead of soda ash. The crystals were large and dissolved more slowly. The acid corroded the steel reactor, the dissolved iron catalyzed exothermic side reactions, and a runaway reaction ruptured the vessel.[2]

*References*

1. Bretherick, L. 1995. In Endothermic compounds. *Bretherick's Handbook of Reactive Chemical Hazards,* 5th ed. Oxford, England: Butterworth-Heinemann.
2. Bretherick, L. 1987. In *International Symposium on Preventing Major Chemical Accidents,* Washington, DC, 3–5 February 1987, p. 4.1. New York: American Institute of Chemical Engineers.

# MYTH 41

## Covering a spillage with foam will reduce evaporation

Water draining from the foam may increase the rate of evaporation from a cold liquid such as chlorine or LPG. To estimate the evaporation produced in this way, Harris[1] assumes that a quarter of the water in the foam will drain in the first 15 minutes and quotes a specimen calculation.

The higher the expansion of the foam, the less the drainage, but high-expansion foams are light and may be blown away by the wind.

If covering a spillage with foam will increase evaporation, then Harris suggests that it be covered with a plastic sheet. A chlorine spillage could be covered by someone wearing breathing apparatus, but I would not like to ask anyone to spread a plastic sheet over an LPG spillage.

*Reference*

1. Harris, N.C. 1987. In *International Symposium on Preventing Major Chemical Accidents,* Washington, DC, 3-5 February 1987, p. 3.139. New York: American Institute of Chemical Engineers.


## MYTH 42

### It is better to fix a problem late than never

To fix a problem late rather than never is not always the best policy. Suppose one liquid is added slowly to another while the mixture is stirred or circulated. If the stirrer or circulating pump stops, the two liquids may form separate layers. If mixing is then started again, the heat of reaction may be more than the cooling system can handle, and a runaway reaction may occur.[1,2]

A pipeline was wrapped after corrosion started. This made the corrosion worse, and the pipeline failed catastrophically. If it had been left unwrapped, it would probably have leaked before failure.

*References*

1. Kletz, T.A. 1994. *What Went Wrong—Case Histories of Process Plant Disasters,* 3rd ed., Section 3.2.8. Houston, TX: Gulf Publishing.
2. *Loss Prevention Bulletin,* No 029, Oct 1981, p. 124; and No 078, Dec 1987, p. 26.


## MYTH 43

### Grounding clips should be fitted across flanged joints

Grounding clips are installed to ensure that the equipment is grounded, so that a charge of static electricity will not accumulate on it.

Certainly, when flammable liquids or gases are handled, all conducting equipment should be grounded, but it is not necessary to install grounding clips across joints. The bolts used to fasten the flanges together make enough contact to ensure a ground connection, as anyone can easily check for themselves with a resistance meter. To prevent a charge of static electricity from accumulating, the resistance to ground should be less than 1 M$\Omega$, but to allow a lightning discharge to flow to ground, the resistance should be less than 7 $\Omega$.

If a section of metal pipe is connected to the main pipe work by a length of glass, plastic pipe, or nonconducting hose, then it should be deliberately grounded. If it is not, a charge of static electricity may collect on the isolated metal and then flash to ground. Reference 1 describes a dust explosion that started in this way and killed a man.

*Reference*

1. Kletz, T.A. 1994. *What Went Wrong—Case Histories of Process Plant Disasters,* 3rd ed., Section 15.3. Houston, TX: Gulf Publishing.

# MYTH 44

## Because water is incompressible, hydraulic pressure tests are safe; if the vessel fails, the bits will not fly very far

Hydraulic pressure testing is safer than pneumatic testing, as much less energy is released if the equipment fails. Nevertheless some spectacular failures have occurred during hydraulic tests. In 1965 a large pressure vessel (16 m long by 1.7 m diameter), designed for operation at a gauge pressure of 350 bars, failed during a pressure test at the manufacturers. The failure, which was of the brittle type, occurred at a gauge pressure of 345 bars, and four large pieces were flung from the vessel. One piece weighing 2 tonnes went through the workshop wall and traveled nearly 50 m. Fortunately, there was only one minor casualty. The failure occurred during the winter, and the report recommends that pressure tests be carried out above the ductile-brittle transition temperature for the grade of steel used. It also states that the vessel was stress relieved at too low a temperature.[1]

Another similar failure has been reported more recently; in this case, substandard repairs and modifications were partly responsible.[2]

When carrying out pressure tests, remember that the equipment may fail and to take precautions accordingly. If we were sure that the equipment would not fail, we would not need to test it. Remember also that if the temperature is too low, equipment may fail during pressurization for service.[2] I do not know of any vessels that have failed in this way, but rupture discs have ruptured because they were too cold.

*References*

1. *British Welding Res Assoc Bull* June 1966. 7(6):149.
2. Snyder, P.G. 1988. *Loss Prevention* 7(3):148.

## MYTH 45

### Electric heaters have a high efficiency

It is first necessary to define efficiency. Almost all the electric energy used in an electric heater is converted into heat. However, if the same amount of electricity is used to drive a heat pump, we get more useful heat. The electric heater has a high First Law of Thermodynamics efficiency (almost 100%) but a low Second Law efficiency (about 25% for a domestic heating system); that is, we could get about 4 times more heat if we used the electricity to drive a heat pump.[1]

There is an informative account of heat pumps in a book by John Sumner.[2] Although everyone has a heat pump or two in their kitchen (in the refrigerator and freezer), many engineers are surprisingly ignorant of its capabilities. Sumner quotes a lecturer who said that a heat pump is impossible, as it offends thermo-dynamic laws!

According to Sumner, the electricity industry in the United Kingdom has discouraged the development of heat pumps, as they would reduce the demand for electricity.

*References*

1. Linnhoff, B. 1983. *Chem Engineering Res Design* 61(4):207.
2. Sumner, J.A. 1976. *Domestic Heat Pumps.* Dorchester, England: Prism Press.

## MYTH 46

### Accidents are due to a coincidence of unlikely events

We are often told, after an accident, that it occurred because several safety devices failed simultaneously, an unlikely coincidence that could not reasonably have been foreseen, so it was hardly practicable to prevent it. In fact, what usually happens is that all the safety devices are left for long periods of time in a failed state. When a triggering event occurs, the accident is inevitable. For example, when an oil tanker was struck by lightning, the vapor coming out of the tank vents was set alight, and the flame traveled back along the vent line into the tanks, where an explosion occurred, killing two men. Three independent safety systems were each capable of preventing the explosion:

- nitrogen blanketing, which was not being used
- a flame arrestor, which was fitted incorrectly, with a gap around the edge
- a pressure/vacuum valve, which opens only momentarily to discharge vapor and the flow rate is too high for the flame to travel back, but a by-pass around the valve had been left open.[1]

There was no coincidence of instantaneous events. There were three ongoing unrevealed (or latent) faults, which stayed unrevealed because there were no regular tests or inspections. When lightning occurred, an explosion was inevitable. Reference 2 describes other incidents due to so-called "coincidences."

*References*

1. *Hazardous Cargo Bull* 1983. 4(7):8.
2. Kletz, T.A. 1994. *Learning from Accidents,* 2nd ed., Sections 2.7, 6.4, and 7.8. Oxford, England: Butterworth-Heinemann.

## MYTH 47

### We can save money by reusing old pipe, etc.

If we are altering or extending the plant, it seems sensible to reuse old pipe that shows no signs of corrosion and looks fit for reuse. Unfortunately, the pipe may be affected in ways that are hard to detect even by expert examination. If it has been used hot, it may have used up some of its creep life (see Myth 21) and may fail in service under conditions in which new pipe would not fail. If the old pipe is made from stainless steel and has been in contact with chloride, it may have been affected by stress corrosion cracking. The amount of chloride in town water may be sufficient to cause corrosion. Old pipe should never be reused unless we know its history and can take advice from a materials specialist.

There are many other examples of accidents caused by penny-pinching. The piston of a reciprocating engine was secured to the piston rod by a nut, which was locked in position by a tab washer. When the compressor was overhauled, the tightness of this nut was checked. The tab on the washer had to be knocked down and then knocked up again. This weakened the washer, so that the tab snapped off in service, the nut worked loose, and the piston hit the end of the cylinder, fracturing the piston rod.

In another incident, the load on a 30-tonne hoist slipped, fortunately without injuring anyone. It was then found that a fulcrum pin in the brake mechanism had worked loose, as the split pin holding it in position had fractured and fallen. The bits of the pin were found on the floor.

Split pins and tab washers should not be reused, but should be replaced every time they are disturbed. Perhaps it is not penny-pinching but lack of spares that prevents us from doing so. Perhaps we cannot be bothered to go to the store. Perhaps there are none in the store.

To quote a Chinese proverb, "There is no gain if you go to bed early to save candles and the result is twins."

## MYTH 48

### In an equation, $g$ is a constant to keep the units correct

The acceleration due to gravity, $g$, is actually a variable, since it can be increased by rotation. This is the principle underlying Ramshaw's Higee distillation system. In this (or any process in which liquid and vapor are in contact), equipment distillation takes place in a rotating disc-shaped bed of packing about 1 m radius and a little less than 1 m thick. Liquid is fed to the center of the bed and travels outward; vapor is fed to the circumference and travels inward. The bed is rotated at about the speed of a centrifuge. The acceleration is about 10,000 m/s$^2$, or about a thousand times gravity.[1-3]

The behavior of a packed bed is described by the Sherwood flooding correlation (Figure 48.1). This shows that if $g$ is increased and we wish to remain at the same point on the curve, the gas and vapor flows can be increased. To visualize what happens, think of the increased "gravity" as reducing back-mixing, which in a normal distillation column, limits the degree of separation that is possible.

The advantages of Higee are reduced size, reduced cost, and a lower hold-up of liquids that are often flammable or toxic. Despite these advantages, very few have been used, even in the company (ICI) in which it was invented. Designers are understandably reluctant to invest in new technology that may have unforeseen snags, especially one that introduces moving parts where there were none before. A

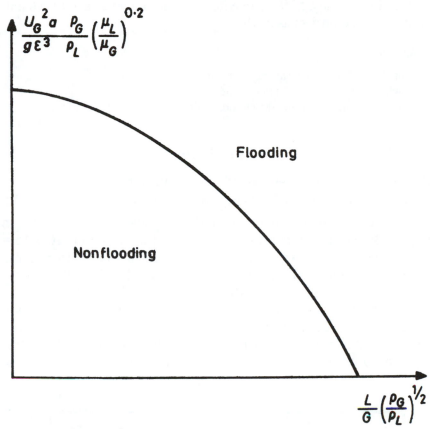

$U_G$ = Gas superficial velocity m/s
$a$ = Packing surface area $m^2/m^3$
$\epsilon$ = Packing voidage
$\rho$ = Density $kg/m^3$
$L$ = Liquid mass flow $kg/m^2 s$
$G$ = Gas mass flow $kg/m^2 s$
$\mu$ = Viscosity kg/ms

Subscripts:
$L$ = Liquid
$G$ = Gas

**FIGURE 48.1** The Sherwood flooding correlation for a packed bed.

cost savings of, say, 50% on one item of equipment is usually a small savings on the project as a whole.

*References*

1. Ramshaw, C. 1983. *Chem Engineer* 389:13.
2. Fowler, R. 1989. *Chem Engineer* 456:35.
3. Literature available from Glitsch Inc., Dallas, TX.

## MYTH 49

### If a liquid of vapor pressure 20 psig is put into a vessel containing nitrogen at 10 psig, the final pressure will be 30 psig

Gauge pressures cannot be added together. For example, in Figure 49.1 the absolute pressure in vessel 1 at the start is 10 + 15 = 25 psi. The absolute pressure in vessel 2 at the start is 20 + 15 = 35 psi. After the transfer, the absolute pressure in vessel 1 is 25 + 35 = 60 psi, and the gauge pressure in vessel 1 is 45 psi.

By way of illustrations consider two boxes standing on the floor, one 25 inches tall and the other 35 inches tall. Draw a line on the wall 15 inches above the floor. The heights of the two boxes above this line are 10 inches and 20 inches. Now stand one box on top of the

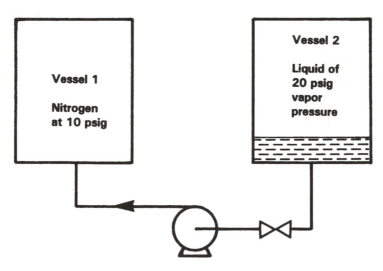

**FIGURE 49.1**  If the liquid in vessel 2 is added to the gas in vessel 1, what is the final pressure in vessel 1?

other. What is the total height above the floor and above the mark on the wall?

## MYTH 50

**If a heat interchanger becomes foul and does not heat the fluid passing through it to the desired temperature, we should increase the surface area**

Obviously, we should, if we can, prevent the fouling of a heat interchanger, for example, by chemical treatment, but this is often not possible. Many engineers would then increase the surface area to compensate for the poor heat transfer. This is learning to live with the problem, rather than curing or preventing it. Linnhoff and co-workers suggest that it may be better to increase the area of another heat exchanger. They discuss an example in detail, involving the fouling of a bottoms cooler on a distillation column, which heats the incoming feed, and they show that it is more economical to provide the extra heat-transfer area elsewhere in the exchanger network. Their solution is not obvious and can be found only by detailed analysis.[1]

They quote an estimate of the money lost by heat exchanger fouling in oil refineries in the western world as $4.5 \times 10^9$ per year (at 1979 prices), so even a small increase in efficiency could produce enormous savings.

*Reference*

1. Kotjabasakis, E., and Linnhoff, B. 1987. *Process Optimisation.* Symposium Series No 100, p. 211. Rugby, England: Institution of Chemical Engineers.

## MYTH 51

### Computers introduce new problems

Software errors are often described as a new sort of error not found before. They do not occur at random but lie in wait like time bombs until a particular combination of conditions arises, perhaps after many years of successful operation, perhaps during a particular alarm condition at a particular stage in a batch. This can also occur with written instructions but with a difference: operators may realize that the instructions are in error (either fundamentally wrong, ambiguous, or ungrammatical) and ignore or correct them, but a computer cannot do so. (In fact, unknown to supervisors and managers, operators sometimes ignore instructions for years.) People, unlike computers, can decipher the meaning of imprecise verbiage; we know what is meant if we are asked to "save soap and waste paper," but computers do not.

Operators can look through instructions in advance and perhaps detect errors; the instructions are usually (though not always) scrutable. Software, however, is incomprehensible to the nonexpert. Even the specification may be hard to read, may not be available to the operating team, or may have been misunderstood by the applications engineer. For example, on one plant the design and operating team told the applications engineer that when an alarm sounded, the computer should hold everything steady until an operator told it to proceed. The applications engineer took this to mean that control valve positions should be held steady, but the operating team really wanted the controlled variables, such as temperature, to be held steady. A human operator could have realized this, whatever the instructions said, and acted accordingly, but a computer could not do so. People can understand what we want them to do; a computer can do only what it is told.

Here are some more examples of familiar errors made more likely by the use of computers.

- It is bad practice, on any plant, to issue blanket instructions such as, "When an alarm sounds, hold everything steady." Each alarm should be considered separately, for each stage of the process, to check that the action is correct.
- Hardware failures are similar to failures on other equipment. They should be foreseen during design and a decision taken whether to install redundancy, warning equipment, or automatic shutdown or just accept the occasional failure. The action we take will depend on the frequency of failure and the seriousness of the consequences. The decision can be checked during hazard and operability studies.
- Many errors occur because operators are misled or confused by poor displays or overloaded by too much information. This can occur on any plant, but there are more opportunities for poor display on a screen. Trends can be seen by glancing at a traditional chart recorder; they may be missed if they are not continuously displayed on a screen and have to be called up.
- Errors occur because operators enter the wrong data. It is easier to do so when entering numbers on a keyboard than when adjusting a setting on an analog instrument.

- Failures have occurred because software was modified, and the change had unforeseen consequences. All modifications to hardware, software, or operating procedures should be formally reviewed and approved before they are authorized (see Myth 25).
- Failures have occurred because operators have made unauthorized changes, acquiring keys or passwords they are not supposed to have, much as operators have always acquired adaptors and tools they were not supposed to use.
- Expert systems may introduce some new problems. If something goes wrong, how is responsibility shared between the expert who provides the knowledge and the "knowledge engineer" who enters it into the expert system? The question is essentially the same as asking how is responsibility shared between the expert who provides information on (say) the course of a reaction and the supervisor who incorporates the information in written instructions. However, the expert can check the written instructions but probably cannot understand the software.

Do not suppose from this list that I am opposed to computer control. Its advantages are enormous, but we should be aware of the problems that can arise so that we can avoid them. Reference 1 gives examples of the errors that can occur on computer-controlled plants and suggests ways of avoiding them. The errors are really human errors, failures to foresee or allow for equipment faults or software errors, failures to understand what the system could or could not do, or failures to realize how people will respond to the displays.

*Reference*

1. Kletz, T.A., Broomfield, E., Chung, P.W.H., and Shen-Orr, C. 1995. *Computer Control and Human Error.* Rugby, England: Institution of Chemical Engineers; and Houston, TX: Gulf Publishing.

# MYTH 52

## Stainless steel is better than carbon steel

We often use stainless steel instead of carbon steel when we need resistance to corrosion, high-temperature strength, or low-temperature ductility, but stainless steel has disadvantages, and there are times when carbon steel or a special type of stainless steel may be better.

- Stainless steels expand much more than carbon steel with rise in temperature. This can set up unforeseen stresses when stainless and carbon steel pipes are welded together. If stainless steel bolts are used with carbon steel flanges, to prevent corrosion of the threads, the bolts may become so loose that the flanges leak.
- If stainless and carbon steels are in contact with the same liquid, they may form an electrochemical cell, and the carbon steel may corrode.
- Stress corrosion cracking can occur in stainless steels that have been in contact with chloride or caustic.
- Stainless steels can withstand higher temperatures than carbon steel. However, they get hotter in service than carbon steel for two reasons: they have a lower thermal conductivity and so lose less heat by conduction, and they are brighter and so lose less heat by radiation. The ability to withstand higher temperatures is therefore not as big an advantage as it appears to be at first sight.
- Stainless steel is difficult to drill or machine, as it hardens ahead of the tool unless special precautions are taken. In addition, the islands of hardened metal can produce stresses.

Underwood[1] describes incidents caused by all these features of stainless steels.

*Reference*

1. Underwood, A.C. 1992. *Chem Engineering Progr* 88(6):69.


## MYTH 53

### The most valuable writings on safety are codes and standards

Codes and standards are certainly valuable, indeed essential, but they hardly grab our attention and get read at once. Unless we read them and act on them, they are ineffective. In contrast, accident reports do grab our attention, do get read, and often spur us into action. If we see a code of practice on the control of modifications, for example, we may put it aside to read when we have time. If we read it, we may agree with it but may soon forget it and do nothing.

In contrast, the report on Flixborough (see Myth 25), when it came out, made many ask if they were doing enough to prevent similar incidents on their own plants. Even today, accounts of Flixborough have an impact and encourage us to read procedures for modification control.

Accident reports are not just bait to get us to read codes and standards. They tell us the important thing: what really happened. We may disagree with a code, but we cannot deny the facts in an accident report. They happened and may happen again. If we do not want to follow the author's recommendations, we realize we have to do something else instead.

Unfortunately, accident reports are forgotten after awhile, even on the plant where they occurred, as people move on and take their memories with them. We should recirculate old reports every few years and provide accident reminders in other ways (see Myth M10).

## MYTH 54

### Keywords tell us the subject of a paper

A paper carried the title "Manganese Mill Dust Explosion."[1] People interested in dust explosions, especially those involving metals, will want to read the paper. They may have found it from one of the abstracting publications or by a computer search; the keywords were "explosion," "dust," and "manganese." Most other readers will have looked at the title and decided it was not for them. If so, they will have missed several important messages that were included in the paper but have little to do with dust or explosions.

The paper describes an incident when a screen became clogged with manganese dust. Before clearing it, the work crew isolated the power supply by means of an emergency switch. Unknown to them, this switch also isolated the power supply to the nitrogen blanketing equipment and the equipment that measured the oxygen content and sounded an alarm if it became too high. Air leaked into other sections of the plant, was not detected, and an explosion occurred.

A general message therefore is, When you isolate equipment for repair, make sure you do not isolate other equipment, especially safety equipment that is still needed. A second general message is, Do operators and maintenance workers understand how equipment works? They may know how each piece of equipment works, but it is also important to know how different items of equipment are linked together, that is, how the system works.

Two more general messages, more familiar ones, also came out of the report. Air entered the equipment because a blind flange had not been inserted, and the screen became clogged because it was finer than usual. Changing the screen size was a modification, and its consequences should have been considered before its use was approved (see Myth 25).

It is easy to find references to *manganese* or *dust explosions* in data bases. If we look for concepts such as *inadvertent isolation, knowledge of equipment, isolation for maintenance, or modification,* we may find little or nothing and may miss important papers. When preparing abstracts, indices, or lists of keywords or when carrying out searches, we often miss concepts, for two reasons:

- Abstract words are imprecise. Manganese has no other name, but *isolation for maintenance* might be called *preparation for maintenance, modification* might be called *change,* and so on.
- The conceptual message or idea is usually less obvious than other messages and may be missed by an indexer or abstracter who cannot be an expert in everything. The word we are looking for may not even be used in the paper; a paper on an accident caused by a modification may not even use this term.

Information scientists should, I suggest, give some thought to ways of overcoming these problems. Perhaps authors should give more emphasis to the general messages in their papers, but they need guidance from editors.

*Reference*

1. Senecal, J.A. 1991. *Loss Prevention Process Industries* 4(5):332.

## MYTH 55

### Supercritical fluids are a distinct form of matter, with properties intermediate between those of liquids and gases

In recent years, fluids above their critical temperatures and at high pressures have been increasingly used as solvents in commercial applications. For example, supercritical carbon dioxide is widely used for removing caffeine from coffee, and oxygen in supercritical water can be used to destroy organic wastes without producing the harmful by-products that may accompany incineration.

There have been a number of articles describing a supercritical fluid as a distinct form of matter, neither liquid nor gas, but something in between. For example, one article said, "Above the critical temperature you don't have liquids and gases. You get something that varies its properties continuously from gas-like to liquid-like." Further on, it said, "More recently fluids have been exploited above their critical temperatures."

The author had apparently overlooked the fact that the air around us is above its critical temperature and has been exploited by living organisms for hundreds of millions of years and by mankind since humans learned to exploit fire.

The difference between a liquid and a gas is that a liquid assumes the shape of its container but retains its own volume. A gas assumes the shape and volume of its container. Splashing ceases above the critical point. There are no intermediate forms. If we compress air, or any other gas, above its critical point, its density increases and may become as high as that of a liquid, but it is still a gas and occupies the full volume of its container. There is no pressure at which we can say that it ceases to be a gas and becomes more like a liquid.

In the language of the *Shorter Oxford Dictionary,* a gas is a "completely elastic fluid" (used in this sense since 1779), while a liquid is a substance "in which . . . particles move freely over each other (so that its masses have no determinate shape), but do not tend to separate as do those of a gas."[1]

If supercritical fluids are gases, how can other substances dissolve in them? It is a matter of definition; it is gas mixing rather than solution as normally understood. When a liquid or solid evaporates and mixes completely with a gas at low pressure, for example, when water vapor mixes with air, we usually describe the process as gas mixing or vapor mixing. If the density of the gas is high, similar to that of a liquid, we often describe the process as dissolving, but there is no pressure at which the process changes or at which a change in nomenclature is justified.

Why are supercritical gases such good solvents? Think of the phenomenon as gas mixing rather than dissolving, and the answer is clear. Consider a supercritical gas such as air at ambient (or any other) pressure. Another gas will mix with it in any proportion. Vapors (that is, liquids below their critical temperatures) will mix with it to a certain extent, but once this extent (defined by the vapor pressure) is exceeded, the vapor forms a separate liquid layer. Salts, which have a very low vapor pressure, hardly mix at all.

In Figure 55.1 the triple point is the temperature and pressure at which solid, liquid, and gas can coexist. The line joining it to the critical point gives the vapor pressure of the liquid. As we move along this line, the temperature rises, the liquid expands, and its density becomes less; at the same time, the density of the vapor rises. The two become equal at the critical point. "At temperatures above this point no pressure, however great, can cause the formation of the

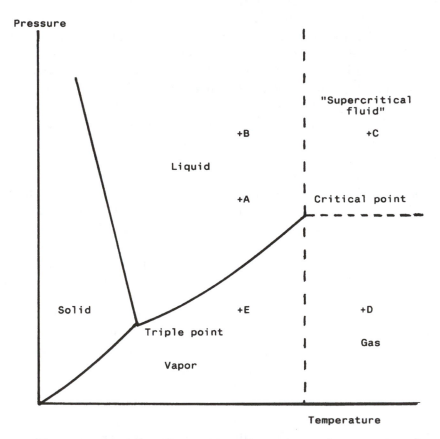

**FIGURE 55.1**  A typical phase diagram. Gases at temperatures and pressures above those at the critical point are often called supercritical fluids. Changes in name occur on passing through the dotted lines, but there is no abrupt change in properties.

liquid phase; at temperatures above the critical point the vapor becomes a gas."[2]

The latent heat also decreases as we move toward the critical point, where it becomes zero.

The term supercritical fluid, a current expression, refers to the area in the right-hand top corner of Figure 55.1, that is, at temperatures and pressures that are both above the critical values.

Imagine a glass vessel filled with liquid at point A, that is, at a pressure slightly above its vapor pressure. Increase the pressure to point B, then raise the temperature to point C. When the liquid passes the critical temperature, it can be described as a gas, but no change will be seen. Next, lower the temperature to point D and then reduce the pressure to point E. We are almost back at the starting point, but the contents have changed from liquid to vapor without any observable change.

*References*

1. *The Shorter Oxford Dictionary.* 1975. Oxford, England: Clarendon Press.
2. Findlay, A. 1938. *The Phase Rule and Its Applications,* 8th ed., p. 17. London: Longmans Green.

# Myths About Management

In management, as elsewhere, procedures may be justified by logic but derive their authority from myth. Myths provide charters for present action.

Origin unknown but based on
the work of Bronislaw Malinowski

The myths about management described in this part have much in common with the technological myths in the previous part; they are partly true and deeply ingrained, but they differ in one important respect. The myths about technology can be disproved by experience and/or by showing that they are incompatible with well-established facts. Experience shows, for example, that pressure vessels exposed to fire are liable to burst and it is well known that metal loses its strength when hot (see the Introduction to Myths About Technology).

In contrast, myths about management are much more a matter of opinion. They are based on experience, but as with all experience of people and of woolly rather than hard subjects, a different interpretation is always possible. For example, despite my contention that discussions convey information and understanding more effectively than lectures or written instructions (see Myth M5), some might argue that any slight gain is not worth the extra cost in time and money. My experience tells me that the gain is substantial; far more is remembered, and it is no good lecturing if your audience does not remember what you say. However, different societies and organizations have different cultures and expectations, and what suits one may not suit another. In this part, therefore, I try to stimulate thought rather than convert readers to my views. If managers do not agree that discussions are the best way of preventing accidents, they should have alternate procedures in mind. Few readers will be confident enough to say that their colleagues are already so well-informed that no action is necessary.

In this respect, some of the myths about technology resemble the management myths, for example, Myth 28, "The public believes technology is making the world a better place." Some people contend that the media and pressure groups do not accurately reflect public opinion.

## MYTH M1

### Engineers should respond to market needs

If our marketing colleagues see a growing market for mousetraps, engineers are asked to design a plant to make mousetraps, or to increase the output from an existing plant. The engineers are also expected to produce the mousetraps as cheaply and efficiently as they can.

However, this is not the whole story. Very often our marketing colleagues and the public do not know what they want until engineers tell them what they can have. We did not know that we wanted home computers and video recorders until we were offered them. Earlier generations did not know they wanted TV sets, radios, telephones, or motor cars. The operating management of British Rail did not know they wanted a 125-mph train until it was offered to them.[1] Engineers should take the initiative and put forward ways of improving the business. As Boeing says, "Planes make markets."

Writing about the history of ICI, which at the time was the United Kingdom's largest chemical company, Carol Kennedy said, "Winnington, where Ludwig Mond had set up his ammonia-soda works in partnership with John Tomlinson Brunner in 1873, was steeped in the Mond Philosophy that research paved the way for industrial progress rather than have it serve needs already perceived by the market."[2] She quoted Ludwig Mond as saying that "the inventor can create new wants," and a former ICI researcher, Derek Birchall, who argued in favor of "science push" rather than "market pull," as saying, "The huge market for polyethylene (a Winnington invention) came about because it was discovered, and was not created because it was needed. . . ." Kennedy added that this view is not popular today and "strikes a distinctly iconoclastic note."[2]

However, chemical engineers are probably less able than other engineers to invent new products because the new products of the process industries are usually invented by chemists. However, if chemical engineers cannot invent new products, they can find better ways of making products: new designs of reactors, separation equipment, heat transfer equipment, etc. A discussion of examples is beyond the scope of this book, but there are some examples of existing and potential inventions in reference 3.

Reactors, separation equipment, and the like are, of course, the products of the process equipment industry, and reference 4 discusses why it has been so poor at innovation. One reason is the low level of employment of first-rate engineers, but university curriculums are also criticized.

If chemical engineers have not always been as good as they might be at innovation, perhaps one reason lies in the next myth.

*References*
1. *Modern Railways* 1984. 41(427):169.
2. Kennedy, C. 1986. *ICI—The Company That Changed Our Lives,* pp. 60, 181. London: Hutchinson.
3. Kletz, T.A. 1991. *Plant Design for Safety—A User-Friendly Approach.* Washington, DC: Taylor & Francis.
4. Solbett, J. 1983. *Process Plant R & D and Innovation.* London: Process Plant Economic Development Committee of NEDO. (For a summary, see *Chem Engineer,* No 394, July 1983, pp. 5, 7.)

## MYTH M2

**The job of the engineer is to answer the questions asked and to solve the problems given**

Engineers are good at answering questions. Tell them your problem and before long you will have many solutions to choose between. We are less good at asking the right questions. We often ask the wrong question. Before answering a question, we should ask ourselves what the questioner really wants to know and if he or she is asking the right question.

Here are some examples.

1. *Question:* How can we reduce the number of fires and explosions at our plants and the damage they cause?

   *Answer:* By detecting leaks automatically, isolating them by means of remotely operated valves, dispersing the leaking material by the use of open construction and steam and water curtains, removing sources of ignition, and finally, if the leak should ignite, by installing fire protection and fire-fighting facilities.

   *Comment:* Much less attention has been given to the question "Why do leaks occur and how can we prevent them?" The usual answer is by good design, construction, operation, and maintenance. I suspect that the most effective way will be by providing better control of construction and better inspection after construction (see Myth 29). More fundamentally, can we use nonflammable or less flammable liquids or use less of the flammable ones or use the flammable ones at lower temperatures and pressures? (See Myth 26 and reference 1).

2. *Question:* How can I improve the control of this process?

   *Answer:* By adding on lots of control instrumentation.

   *Comment:* Here are two examples of ways in which someone has avoided the need for a lot of added-on control equipment.

   - In the NAB process for the manufacture of nitroglycerine, the ratio of the two reactants, acid and glycerine, has to be kept constant. The laws of physics are used instead of a flow ratio controller. The acid flows through a device rather like a laboratory water pump. The acid sucks in the glycerine through a side arm. If the flow of acid changes, the flow of glycerine changes in proportion, without the intervention of control equipment, which might fail or be neglected.[2]

     The "water pump" is the reactor, and the main advantage of this process over earlier ones is that the inventory in the reactor is greatly reduced (see Myth 26). The inherent control is a useful bonus.

   - A difficult control problem, controlling a temperature to within 1°C, was solved by controlling another temperature

that was related to the first one but showed greater variation.[3] Wide temperature swings in a distillation column
were prevented by controlling the temperature on a particular tray.[4]

In other cases it may be possible to modify the process so
that we do not need to control it so accurately.

3. *Question:* Many accidents have occurred because operators
opened the wrong valve or forgot to open a valve. How can
we prevent people from making such a mistake?
*Answer:* By telling operators to be more careful, penalizing
those who make mistakes, etc.
*Comment:* People carrying out a routine task will make occasional mistakes even though they are well trained, well moti-
vated, and physically and mentally capable. Once we realize
this, and if the occasional mistakes cannot be accepted, then
the problem becomes one of removing or reducing the opportunities for error (see Myth 6).

4. *Question:* What research should we do on safety?
*Answer:* Many suggestions have been made and carried
out, some of them quite expensive, for example, large-
scale tests on gas dispersion and unconfined vapor cloud
explosions.
*Comment:* While there is much we would like to know, on
the whole, accidents are not the result of lack of knowledge.
Accidents occur because the knowledge of how to prevent
them, though available, is not known to the people concerned, or they know what to do but lack the will to do a
thorough job.

Myth M10 suggests some ways of increasing people's
knowledge of past accidents and the ways to prevent them.

5. *Question:* How can fire engines get to a fire more quickly?
*Answer:* By providing devices for starting engines more quickly,
opening fire station doors automatically, setting traffic lights in
favor of the fire engine, and so on.
*Comment:* The problem was seen as an engineering one. The
time between someone dialing the emergency number and the
fire station getting the message can be as long as the journey
to the fire. Little or no thought has been given to ways of
speeding up this step.[5]

6.  *Question:* The level of carbon monoxide in a covered car park was too high, so the question was asked, "How can the carbon monoxide be removed?"
    *Answer:* A scheme was prepared for forced ventilation. It was expensive and there was no guarantee that it would be effective.
    *Comment:* When someone asked if the formation of carbon monoxide could be prevented, it was found that the high level was due to cars driving around and around looking for somewhere to park. Traffic lights controlled by traffic counters were installed at the entrance.[6]

7.  *Question:* The Negev Desert in Israel is irrigated with water transported great distances by pipeline. Water has been obtained from deep (over 500 m) wells, but it contains over 500 ppm chloride. Can this be removed so that the water is suitable for irrigation?
    *Answer:* Techniques are available.
    *Comment:* Changes in methods of irrigation (to the drip method) and treatment of the soil have allowed the water to be used without purification.

8.  About 700 BC, King Hezekiah built a tunnel (now open to the public) to bring water from a spring outside the walls of Jerusalem to a point inside the walls. This enabled him to withstand a siege by the Assyrians.[7]

    Archaeologists have been puzzled by the fact that the roof of the tunnel varies in height (from 1.5 m to 5 m). Instead of asking why the roof is so high in places, someone asked why the floor was so low. A possible reason then became apparent: when the tunnel was constructed, the floor was too high in places for the water to flow and had to be lowered.

Why are we so much better at answering questions than at asking the right questions? Is it because we are trained at school and university to answer questions that others have asked? If so, should we be trained to ask questions?

*References*

1. Kletz, T.A. 1991. *Plant Design for Safety—A User-Friendly Approach.* Washington, DC: Taylor & Francis.

2. Bell, N.A.R. 1971. In *Loss Prevention in the Process Industries.* Symposium Series No 34. Rugby, England: Institution of Chemical Engineers.
3. Pickles, R.G. 1971. In *Loss Prevention in the Process Industries.* Symposium Series No 34. Rugby, England: Institution of Chemical Engineers.
4. Horwitz, B.A. 1994. *Chem Engineering Progr* 90(11):62.
5. *Fire Prevention* 1981. No 145, p. 31.
6. *Health and Safety at Work,* Oct 1981, p. 94.
7. II Chron. 32:2–4, 30.

# MYTH M3

## Many questions can be answered by substituting numbers in well-established equations

Below are three cases showing how the wrong answer was obtained in this way. The people who used the equations did so mechanically without comprehending the underlying physical realities. Computers make it particularly easy to do this, but a computer is not necessary, not even a pocket calculator, as the examples will show. We can be slaves of a lesser God.

*Case 1: Fractional Dead Times*

We start with some definitions:

*Hazard rate* is the rate at which hazards occur; for example, the rate at which the pressure in a vessel exceeds the design pressure or the rate at which the level in a vessel gets too high and the vessel overflows.

*A protective system* is a device installed to prevent the hazard from occurring, for example, a relief valve or a high-level trip.

Tests are (or should be) carried out at regular intervals to determine whether or not each protective system is inactive or "dead." It is assumed that if it is found to be dead it is promptly repaired. The time between tests is the *test interval* ($T$).

*Demand rate* ($D$) is the rate at which a protective system is called on to act, for example, the rate at which the pressure rises to the relief valve set pressure or the rate at which a level rises to the set point of the high-level trip. "Demand" is used in the French sense (demander = to ask).

*Failure rate (f)* is the rate at which a protective system develops faults that prevent it from operating.

*Fractional dead time* is the fraction of the time that a protective system is inactive, i.e., the probability that it will fail to operate when required.

If the protective system never failed to operate when required, then the hazard rate would be 0.

If there was no protective system, then the hazard rate would be equal to the demand rate.

Usually the protective system is inactive or dead for a (small) fraction of the time.

A hazard results when a demand occurs during a dead period. Hence

Hazard rate = Demand rate × Fractional dead time

To derive the fractional dead time, we assume that failure is random. It will therefore occur, on average, halfway between tests, and the protective system will be dead for half the test interval ($\frac{1}{2}$ *T*). Failure occurs on average after an interval of 1/*f*, so the fractional dead time is $\frac{1}{2}$ *fT*, and the hazard rate $\frac{1}{2}$ *DfT*.

An example may make this clearer. Consider a high-level trip for which

the failure rate *f* = 0.5/year (once in 2 years)
the test interval *T* = 0.1 year (5 weeks)
the demand rate *D* = 1/year

Failure will occur on average halfway between tests, so the trip will be dead for 2.5 weeks ($\frac{1}{2}$ *T*) every 2 years (1/*f*). The fractional dead time is therefore 2.5/(2 × 52) = 0.024, and the hazard rate is 1 × 0.024 = 0.024/year, or once in 42 years.

Now consider another example. Let

the failure rate *f* = 0.5/year (as above)
the test interval *T* = 0.1/year (as above)
the demand rate *D* = 100/year

Substituting in the formula (hazard rate) = $\frac{1}{2}$ *DfT*, we get

$$\text{Hazard rate} = \tfrac{1}{2} \times 100 \times 0.5 \times 0.1 = 2.5/\text{year}$$

However, this answer is nonsensical. The trip will fail once every 2 years. This failure will probably be followed by a demand rather than a test (as there are 10 demands but only 1 test in each 5-week period), the vessel will overflow, and the fault will be discovered and repaired. The hazard rate will be almost the same as the failure rate, 0.5/year. Thus 2.5/year would be the correct answer if, when the vessel overflowed, we allowed it to overflow twice per week until the next test was due.

If you find this hard to follow, consider the brakes on a car. Let

failure rate $f = 0.1/\text{year}$ (once in 10 years)
test interval $T = 1$ year (as required by law)
demand rate $D = 10,000/\text{year}$ (a guess)

Substituting in the formula (hazard rate) $= \tfrac{1}{2} DfT$, we get

$$\text{Hazard rate} = \tfrac{1}{2} \times 10,000 \times 0.1 \times 1 = 500/\text{year}$$

Not even the worst drivers have this many accidents. Clearly, the simple formula (hazard rate) $= \tfrac{1}{2} DfT$, which we derived, does not always apply. By using it mechanically, without thinking of the reality behind it, we got wrong answers.

A more correct formula is

$$\text{Hazard rate} = f(1 - e^{-DT/2})$$

When $DT/2$ is small, this becomes

$$\text{Hazard rate} = \tfrac{1}{2} fDT$$

When $DT/2$ is large, this becomes

$$\text{Hazard rate} = f$$

For a table comparing these equations, see reference 1.

*Case 2: Factorial Costing*

It was decided during the design of a new plant (total cost about $10 million) that the control building should be strengthened so that it could withstand the effects of an unconfined vapor cloud explosion.

The project manager later reported that the increase in cost was $100,000, a sum so large that some people questioned the wisdom of the decision, and an investigation was requested.

The civil engineer who had designed the control building said that the increase in cost was only $30,000, about 14% of the original building cost. The project manager had multiplied this figure by a factor of 3.3 to arrive at an overall cost increase of $100,000.

Factorial costing is a well-recognized method of estimating plant costs. The cost of the main plant items (reactors, distillation columns, heat exchangers, etc.) is multiplied by a factor, usually in the range of 3-6, depending on the type of plant, to allow for land, foundations, structures, piping, electrical equipment, instruments, ancillary buildings, and utilities as well as design and construction. The control building was considered a main plant item, and a factor of 3.3 was the achieved figure for that particular plant.

However, increasing the strength of the control room may have added a little to design and construction costs but would not have increased the other items listed above. The figure of $100,000 was nonsensical. It had been obtained by using a formula mechanically without thinking of the reality behind it or asking if it applied in this case.

### Case 3: Dates

Suppose a pole is inserted in the ground so that the length below ground is 1000 mm and the length above ground is 2000 mm. The length of the pole is 1000 + 2000 = 3000 mm.

Suppose an event occurred on June 1 (or any other date) in the year 1000 B.C. and is expected to occur again on the same date in the year 2000 A.D. How many years apart are the two events?

Many people would use the same calculation and say 3000 years. The correct answer is 2999 years, as there was no year 0.

### Conclusions

My examples have been relatively simple ones, but they are supported by Linnhoff, who wrote, ". . . the basic understanding of important principles . . . can be a far more powerful asset in design than systematic methods. Systematic methods exclude the engineer from the design task. Above all, design is a creative process and to exclude the engineer from it must be a mistake."[2] Linnhoff goes on to say that he has shifted his emphasis from the "invention of methods" to the "demonstration of principles in use."

Just as systematic methods exclude the engineer from the design task, overly formal management systems may exclude the manager from the management task. For example, if a company has a rigid system for investigating accidents and informing people of the results (or fixing the rate of pay for a job), managers may be lured into a belief that all they need do is follow the system. In fact, each accident (or job) is different, and a flexible approach is needed.[3] The system inhibits flexibility and innovation. Similarly, a system of detailed safety laws and regulations encourages managers to think that all they need do is follow the rules (see Myth 51).

According to Ernest Siddall, good engineering system design can be hindered by formal methods. If the designs are good, "it will be because the engineers concerned will have found ways to circumvent the impediments and fill in the deficiencies of the present formal system of programming. . ."[4]

There has been much interest recently in safety and environmental management systems, and many have been described in journals and at conferences. While they are desirable, in fact, necessary, they are not sufficient. All they can do is harness people's knowledge and experience in a systematic way and thus reduce the chance that something will be overlooked. If the staff lacks knowledge and experience, the system is an empty shell. Unfortunately, this is not always realized, and some companies seem to think that if they follow a management system, all will be well.

Perhaps we can make a further analogy. In life, there are (according to Isaiah Berlin[5]) the "hedgehogs," who have an all-embracing system of belief and conduct, and the "foxes," who have no such unifying vision. When the hedgehog is challenged, he rolls up into a ball and displays his spikes—the rules that tell him what to do—the same strategy for all occasions. The fox is more flexible; he will vary his strategy to suit the occasion and will accept and seek to justify some degree of contradiction in aims and methods. The hedgehog finds old answers to new questions; the fox, new answers to old questions.

In safety, hedgehogs like a set of detailed regulations covering all eventualities; foxes prefer to do what is reasonably practicable (see Myth 51). The hedgehog and fox metaphor is taken from the Greek poet Archilochus, who wrote, "The fox knows many things but the hedgehog knows one big thing."

*References*

1. Kletz, T.A. 1992. *Hazop and Hazan—Identifying and Assessing Process Industry Hazards,* 3rd ed. Rugby, England: Institution of Chemical Engineers.
2. Linnhoff, B. 1983. *Chem Engineering Res Design* 61(1):207.
3. Kletz, T.A. 1994. *Learning from Accidents,* 2nd ed. Oxford, England: Butterworth-Heinemann.
4. Siddall, E. 1994. *Highly Reliable Systems—Hardwired or Computerized.* Waterloo, Ontario: Department of Systems Design Engineering, University of Waterloo.
5. Berlin, I. 1953. *The Hedgehog and the Fox,* especially pp. 1, 80. London: Weidenfeld and Nicholson.

# MYTH M4

## Accidents can be prevented by following detailed rules and regulations

The United States and United Kingdom take different approaches. In the United States, and many other countries, the government writes a book of regulations that looks like a telephone directory, but is less interesting to read. These regulations have to be followed to the letter, whether or not they will actually make the plant safer or are appropriate to our particular problems.

In contrast, in the United Kingdom, under the Health and Safety at Work Act (1974), still in force and the main item of safety legislation, there is a general obligation on employers to provide a safe plant or system of work and adequate instruction, training, and supervision. It is up to the employer to decide which designs and procedures make for a safe plant, but the factory inspector can disagree and, if necessary, issue an Improvement or Prohibition Notice.

If there is a generally accepted code of practice, failure to follow it is prima facie evidence of an unsafe plant or system of work. However, the employer can argue that the code is inapplicable in his case or that he is doing something as safe or safer.

The points in favor of the UK system are as follows.

1.  Codes are more flexible than regulations. They can be changed when new problems arise or new solutions are found to old problems, and in any case, as already stated, they do not have to be obeyed to the letter. Changing regulations, however, takes a long time (see example below).

2. Employers cannot take shelter behind the regulations. They cannot say (as I have heard people say in some countries), "My plant must be safe because I have followed all the regulations." (There was something of this attitude at Three Mile Island.[1])
-> Employers must look out for and control any hazards missed by the regulations.
3. It is impracticable to write detailed regulations for a complex, rapidly changing technology.
4. The UK code system is a more powerful weapon in the hands of a factory inspector than a volume of detailed regulations. Under a regulatory system, the inspector has to show that a particular regulation has been broken. If not, he is powerless. Under the UK system it is sufficient to show that the employer has not provided a safe plant or system of work, so far as is reasonably practicable.

   Some people think that under the UK system the employer has a soft option. However, the pressures are actually greater than under the old (pre-1974) system.

Here are two examples of nonsensical occurrences that can arise under a system of detailed regulations that have to be observed to the letter.

1. Under the UK Factories Act (1961) and earlier legislation, steam boilers had to be inspected at intervals not greater than 26 months. Those who drew up the regulations had fired boilers in mind, but the regulations applied to all boilers, including waste heat boilers on oil and chemical plants, which may be subjected to much less severe conditions than fired boilers. Shutting down plants in order to carry out the statutory inspection of a waste heat boiler was often expensive and unnecessary.

   The factory inspectorate had power to grant exemptions, but the procedure was cumbersome.

   Other plant vessels (except air receivers and steam receivers) did not have to be inspected under the Factories Act. Regular inspection is now required by the Health and Safety at Work Act (1974) and subordinate legislation, but an appropriate inspection frequency can be chosen.
2. Under the UK Petroleum (Consolidation) Act (1928), tank trucks carrying petroleum spirit had to be emptied by gravity.

They could not be emptied by blowing the contents out with compressed nitrogen.

Expecting a change in the law, a company ordered some tanks that were to be emptied in this way, for carrying petroleum spirit between two factories. In order to withstand the nitrogen pressure, they had to be stronger than ordinary tanks and were therefore less likely to be damaged in an accident. The change in the law was postponed for several years, and the company had to decide what to do with the tanks. The Health and Safety Executive publicly said that they would turn a blind eye to their use but that it was possible, though unlikely, that the police might prosecute for a technical breach of the law, if one of the vehicles was involved in an accident. The company lawyer advised that the vehicles not be used.

*Reference*

1. Kletz, T.A. 1982. *Hydrocarbon Processing* 61(6):187.

*Further Reading*

*Report of the Committee on Health and Safety at Work (Chairman, Lord Robens).* 1972. London: Her Majesty's Stationery Office. (For the background to the Health and Safety at Work Act.)

*Her Majesty's Inspectors of Factories 1833–1983: Essays to Commemorate 150 Years of Health and Safety Inspection.* 1983. London: Her Majesty's Stationery Office. (Gives a good account of the UK system of factory inspection, of which the Health and Safety at Work Act is only a late development.)

## MYTH M5

### The best way of conveying information to people is to tell them

It seems to be true that talking is a more effective way of getting a message across to other people than writing to them, although we may have to follow up with the written word to reinforce, to explain details, and to provide a source for reference. Less important, the written word may provide evidence that we conveyed the message. Talking to people singly or in small groups is, of course, more effective than talking to large groups because feedback is easier—we know whether the message is received and what reaction it causes.

If we have to convey messages that people want to receive ("where to get free beer," for example), almost all methods of communication are effective. However, if there is some resistance to the message, as there often is when we are making recommendations to increase safety, for example, then we should choose the most effective method of communication: discussion. The following technique has been found particularly effective for putting across safety messages.

An accident is illustrated by a slide. The discussion leader explains very briefly what happened. The group then questions the discussion leader to establish the rest of the facts and say why they think the incident occurred. They then say what *they think* should be done to prevent the incident from happening again, not only in the plant where it actually occurred but in other plants as well.

Such discussions take longer than a lecture, but more is remembered and people are more committed to the conclusions because they have not been told what to do but have worked it out for themselves.[1]

If possible, the incidents discussed should have occurred in the same company or factory, but if suitable incidents are not available, sets of notes and slides can be obtained from the UK Institution of Chemical Engineers.[2]

The best size for a discussion group is 12–20. If fewer than 12 are present, the group may not be "critical" (in the atomic energy sense) and discussion may not take off. If more than 20 are present, the quieter members may not be able to contribute.

Similarly, if operating or safety instructions are being revised, rather than just issuing the new instructions, it is better to explain them to those who will have to carry them out, listen to their comments, and answer their questions. It may be that what is being proposed is in some respects impracticable.

An argument against discussions is given in one of Harry Kemelman's novels.[3] A group of students suggests a discussion instead of a lecture, and the lecturer asks if they hope to achieve knowledge by combining ignorance.

However, in many situations, people have the knowledge but do not seem able to draw on it. The stimulus of a discussion often helps to break down the barriers between different parts of the mind. Also, while we cannot achieve knowledge by combining ignorance, we may achieve knowledge by combining our individual bits of knowledge.

*References*

1. Kletz, T.A. 1993. *Lessons from Disaster—How Organisations Have No Memory and Accidents Recur.* Rugby, England: Institution of Chemical Engineers; and Houston, TX: Gulf Publishing.
2. *Safety Training Packages.* Rugby, England: Institution of Chemical Engineers. The titles include
   *Hazards of Over- and Under-pressuring of Vessels*
   *Preparation for Maintenance*
   *Fires and Explosions*
   *Human Error*
   *Modifications: The Management of Change*
3. Kemelman, H. 1976. *Tuesday the Rabbi Saw Red,* p. 49. Greenwich, CT: Fawcett Publications.

## MYTH M6

## We can rely on the advice of the accountant where money is concerned

I describe below cases where the wrong technical decisions have been made as the result of following the company's financial procedures.

### Capital and Revenue

In many companies it was at one time the practice (in some it still is) to treat capital and revenue as if they were different commodities that could not be combined. As a result, designs were chosen that were cheap in capital but expensive to maintain. The 1960s were the time of the "minimum capital cost" philosophy. Since then, industry has on the whole become more realistic—discounted cash flows have been widely used—but errors still occur.

For example,

1. A colleague of mine once watched a new plant being built from his office window. He calculated that by the time the plant was complete the cost of hiring the scaffolding around the distillation columns would have paid for a permanent structure, without taking into account the cost of hiring and erecting scaffolding for shutdowns during the life of the plant.

2.  When equipment has to be isolated for maintenance, the most effective method of isolation is a slip-plate (blind or spade) [Figure M6.1(a)]. A variation on the slip-plate is the figure-8 (spectacle) plate [Figure M6.1(b)]. In one position it allows flow; when it is turned, it acts as a slip-plate. It is also safer, as we can see at a glance whether the line is slip-plated or open.

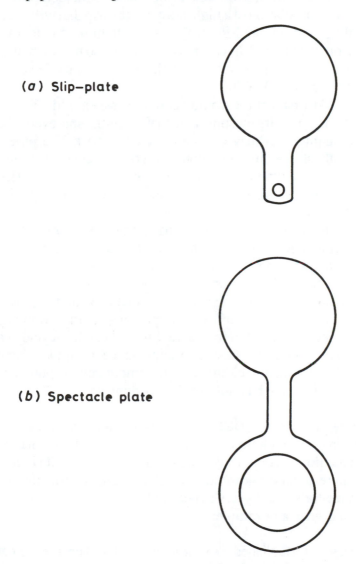

(*a*) Slip—plate

(*b*) Spectacle plate

**FIGURE M6.1**   Two methods of isolating pipework.

The disadvantage of the figure-8 plate is its extra cost. However, it is fixed in position. In contrast, slip-plates disappear when not in use, and more are forever having to be ordered. Figure-8 plates may be cheaper in the long run, especially when isolation for repair is frequent.

3. Suppose that we wish to prevent a pump from overheating if a valve in the delivery line is closed. Two methods are considered: installation of a high-temperature trip that will shut down the pump [Figure M6.2(a)] and installation of a kickback line from the pump delivery back to storage, so that there is always a small flow [Figure M6.2(b)]. Let us suppose that (a) costs $4000 and (b) $6000.

   At first sight it would seem that we should choose a, but the trip requires testing and maintaining, and even after discounting, this may double the capital cost. The pipeline, in contrast, has little maintenance cost attached to it, though it does use a little power as the pump continues to run. (In many real-life cases the pipeline would be cheaper in capital cost as well as revenue cost.)

4. In the United Kingdom, during the 1960s and 1970s, the quality and durability of most of the shopping center developments were poor, as the developers' sole concern was initial cost and short-term profits. Whole-life costs were ignored.[1]

5. In 1985, following the realization that 80% of delays were due to the failure of 20% of components, British Railways announced that in the future they would use life-cycle costing,[2] that is, take the cost of maintenance or replacement into account when deciding which components to purchase. Apparently, they had not previously done so.

Discounted cash flow (DCF) is widely recommended as a way of overcoming the dichotomy between capital and revenue but, as Malpas has pointed out,[3] if we are not careful, it can stifle innovation. When interest and thus discount rates are high, benefits that will not materialize for several years appear to have little value.

Malpas gave an example:

Process A (established): Assume raw material costs are steady for 3 years and then rise 10%/year.

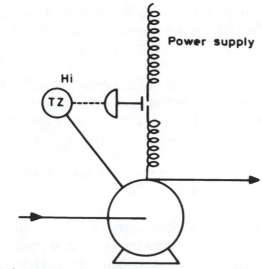

**Power supply**

**Hi**

**TZ**

**(a)** High—temperature trip

**Suction vessel**

**Restriction orifice plate**

**(b)** Kickback

**FIGURE M6.2**   Two methods of preventing a pump from overheating.

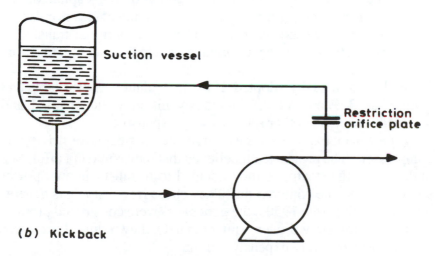

Process B (new): Assume raw material costs are initially double those for Process A but fall rapidly, becoming equal in 5 years and then rising at 5%/year.

After 6 years, the cash flows are the same, and from then on are much in favor of B. But DCF calculations favor A.

The DCF figures do show that Process B is favored in the long run, but companies are then reluctant to invest in Project B.

*Notional Capital*

On a large integrated site a central organization supplied steam and other services to a large number of plants operated by other divisions of the same company. In this situation there are several possible ways of charging the user plants for services:

1. Market value—what it would cost the consuming plants to buy the services elsewhere (if they could) or manufacture them themselves.
2. Cost of production plus a reasonable return on capital (say, 10–15%). Because there was a reasonably assured market for the services, this return would be less than expected for a new product that carried a good deal of market risk.
3. The consuming plants carry a share of the capital of the services plants, proportionate to the amount of steam and other services they expect to use. There is a notional transfer of capital from the services organization to the user organization.

Method 3 was actually adopted. If the user plant was profitable and earned, say, 30% on capital, it had to pay this on services capital, and this made steam and other services very expensive.

A new plant expected to be a large user of steam, for driving its compressors. The project team believed that they could not afford the high cost of the steam, so they decided to install their own power source to drive the compressors. They chose free-piston gas generators. At this time (the 1950s) the generators were considered promising if not completely proven. Unfortunately, they required extensive maintenance and were expensive to run.

If the services organization had been less "greedy" and had been content with a more modest return on capital, the consuming organization would have used steam. The profits of the organization as a whole would have been greater.

*Allocation of Overheads*

In the Mexican branch of an international chemical company, overheads were allocated in proportion to the number of invoices issued. This imposed a ludicrous penalty on a business (pharmaceuticals) that sold small quantities to large numbers of customers. It was found to be easier to transfer the business to an agent than to change the accounting system.

### Costing Similar Products Together

Sir John Harvey-Jones, former chairman of ICI, described a group of products that performed well for many years[4]:

> The cost accountancy criteria had been set up many years before, and had in their day been considered to be outstanding examples of a sensible way of allocating costs and overheads across an enormous range of associated products. The system appeared to work well for many years, but the business met hard times and many changes had to be made ... we put in a new accountant from a totally different background, who examined the whole basis of our costing, *ab initio*. Within a remarkably short space of time we found, as is almost always the case, that what we thought was a universally bad business contained some products which were in reality extremely good, and a minority which were extremely poor. The minority had concealed the performance of the good but because we had looked at the whole lot together we had failed to appreciate this. This may seem a very elementary mistake for a large and sophisticated organization to make, but I would caution against too happy a view that it could not happen in your organization. ...

### One Industry Makes the Products of Another

This example is more concerned with national rather than company accounting. Methods were developed for manufacturing single-cell protein from hydrocarbons or methanol.[5] The product was intended primarily for sale as a constituent of animal feedstuffs as an alternative to soya bean meal.

The raw materials for single-cell protein were the products of the oil and petrochemical industries, and manufacture was carried out by oil and petrochemical companies, so the projects were evaluated as typical oil and petrochemical projects and subjected to rigorous investment analysis. In most countries, agricultural projects are assessed somewhat differently. Subsidies or guaranteed prices are often available to encourage domestic food production or even for sentimental reasons ("farming is part of our heritage"). Should single-cell protein manufacture be assessed as a farming project or as a typical manufacturing project?

If new methods of making agricultural products are assessed as manufacturing projects, can they ever compete in the marketplace?

### Depreciation

In 1989, when the UK government was considering the privatization of electricity production, the public was surprised to find that the cost of nuclear power was over twice previous estimates. Much of the increase was due to changes in accounting. In 1987 the cost of producing electricity in a new pressurized water reactor was said to be 3.09 p/kWh (1 p ≈ 1.5 cents), assuming 8% return on capital and 40 years depreciation, as the market was assured. In 1989 the estimate was 6.25 p/kWh. Of the increase, 1.4 p/kWh was due to a change to 10% return and 20 years depreciation, as under privatization the market would no longer be assured. The rest of the increase was due to inflation of 0.33, overheads (not previously included) of 0.79, and technical factors (mainly decommissioning costs) of 0.64.[6,7]

### Afterthought

Early in my career I was asked to make a quick estimate of the cost of manufacturing a particular product. A published figure for the same process was much lower. My boss wrote across the top of a sheet of paper:

Reconciliation of ICI's and ———'s Estimates of the Cost of Making———.

He then went on to show that the two estimates were really the same. He did not change any of my technical figures, such as raw material efficiency or services usage; he did it all by changing accountancy conventions.

### References

1. Best, K. 1992. *Best Endeavours,* p. 119. York, England: Keith Best.
2. Faith, N. 1985. *The Economist* 296:25.
3. Malpas, R. 1983. *Harvard Business Rev,* July–Aug, p. 122.
4. Harvey-Jones, J.H. 1988. *Making It Happen.* London: Collins.
5. Craig, J.B., and Lloyd, D.R. 1984. ICI technology for SCP production from methanol and Its wider application. In *Proceedings of the CHEM-RAWN III Conference* (CHEMical Research Applied to World Needs), June. The Hague, Netherlands: Royal Dutch Chemical Society.
6. Marshall, W. 1990. *Atom* 400:5.
7. Jones, P. 1990. *Atom* 400:34.

## MYTH M7

**Debottlenecking is the best way of increasing the capacity of a plant**

Debottlenecking often gives extra capacity for low capital cost. However, as Malpas has pointed out,[1] there are disadvantages. Excessive debottlenecking draws scarce technical resources away from innovation and delays the date when brand new plants are built. It is these new plants, if they utilize new technology, that can make the greatest contribution to productivity, energy savings, environmental improvement, and safety.

The *extra* capacity obtained by debottlenecking may be bought cheaply, but the effect on total costs is often small compared with the economies possible by introducing new technology. If output is limited by a few pumps or heat exchangers, obviously we

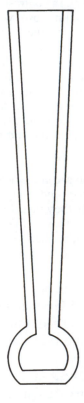

**FIGURE M7.1**   Do not try to debottleneck this bottle!

replace or supplement them, but some bottles are nearly all neck (Figure M7.1).

*Reference*

1. Malpas, R. 1983. *Harvard Business Rev,* July–Aug, p. 122.

## MYTH M8

### The boss does not need to get involved in the details

Many a young engineer, after promotion, decides to stand back, take a broad view, and leave the details to his subordinates. There is no shortage of books on management recommending this practice. R.V. Jones espouses the opposite view[1]:

> The experience brought home to me what my real strength had been at the earlier meetings. It was that, in contrast to everyone else sitting round the Cabinet table, I had done all my own work for myself and had forged out every link in the chain of evidence, so that I knew exactly what its strength was. Everyone else, in their more elevated positions, had had to be briefed at the last moment, as I myself had had to be on this occasion. And even with Charles Frank's understanding and skill, and even though I had been away from the work for only a week, I felt that there was too much sloppiness in my knowledge for me to pronounce positively on the various possibilities. On a previous occasion I had quoted Palmerston's statement of 1838 to Queen Victoria, and now I even more appreciated its force:

> In England, the Ministers who are at the heads of the several departments of the State are liable any day and every day to defend themselves in Parliament; in order to do this they must be minutely acquainted with all the details of the business of their offices, and the only way of being constantly armed with such information is to conduct and direct those details themselves.

Professor Jones also describes how two physicists were assigned to work on the characteristics of German magnetic mines. One stuck to his desk, working on reports from minesweepers; the other went to sea and soon realized that the minesweepers' reports were hopelessly inaccurate.

We need to immerse ourselves in the details—otherwise, we do not understand all aspects of a particular process—but we should then stand back and view the wider scene. This is not easy.

To quote Sir John Kendrew, "To turn out good scientists you have to keep their noses to the grindstone quite a lot, but you must at the same time try and stimulate their interests on a broad front, so that, when they do reach the stage of being independent creative people, they will go off into new fields; and this is very difficult."[2]

Summarizing the lessons of Three Mile Island, J.F. Ahearne wrote, "If the boss is not concerned about details, his subordinates will also not consider them important. . . . It is hard, monotonous, and onerous to pay attention to details; most managers would rather focus on lofty policy matters. But when details are ignored, the project fails. . . . Most managers avoid keeping up with details; instead they create 'management information systems.'"[3]

*References*

1. Jones, R.V. 1978. *Most Secret War,* p. 353. London: Hamilton.
2. Kendrew, J. 1974. *Chem* Britain 10(11):443.
3. Ahearne, J.F. 1986. Three Mile Island and Bhopal: Lessons learned and not learned. In *Hazards: Technology and Fairness,* p. 197. Washington, DC: National Academy Press.

## MYTH M9

### Spending money to prevent accidents will reduce the injury and damage they cause

In the United States, flood damage has increased as expenditure on flood control has increased.

The reason for this occurrence is that when floods were frequent, there was little development in areas subject to flooding. When flood prevention measures were introduced, development took place in these areas. When an exceptional flood overwhelmed the flood prevention measures (or the measures failed to achieve design performance), the damage was much greater than in the days before flood prevention was practiced.[1] This occurred in the Mississippi River basin in 1993.

Efforts were made to reduce the risk of forest fires in the U.S. National Parks. They were at first successful. Fires were fewer and scrub and brushwood flourished. However, when a fire did occur, it was of a size and intensity never experienced before.[1]

A tank was filled once a day with sufficient raw material to last for 24 hours. An operator watched the level and switched off the filling pump when the tank was nearly full. This took place without incident for 5 years, until one day the operator allowed his attention to wander, and the tank was overfilled. The spillage was small because the operator caught it early.

A high-level trip was then fitted to the tank. The designer's intention was that the operator would continue to watch the level and that the trip would take over on the odd occasion when the operator forgot to do so. The chance of the operator and trip failing at the same time was negligible. However, the operator left the control of the level to the trip and continued with other work. The trip failed, as could have been predicted, within 2 years, and the spillage was much greater than before because the operator was not on the job.

On a larger scale, one of the most important factors influencing the probability of death per year is social class. The death count from accidents is small in comparison. In the United Kingdom today the mortality of adult males in social class 5 (unskilled occupations) is 1.8 times that of adult males in social class 1 (professional occupations). For children the ratio is slightly greater.[2,3] Instead of taxing low-income people so that the government can spend money on reducing remote risks, like those from nuclear power stations, would we save more lives if we left money untaxed, or used it to increase the standard of living?[2]

### References

1. Clark, W.C. 1980. In *Societal Risk Assessment,* eds. R.C. Schwing and W.A. Albers, p. 287. New York: Plenum Press.
2. Wildavsky, A. 1980. *The Public Interest,* summer, p. 23.
3. Townsend, P., and Davidson, N. (eds.) 1988. *Inequalities in Health,* Chap. 2. London: Penguin Books.

# MYTH M10

## We need to know what is new

We do need to know what is new, but that should not negate our concern with what is old. In my own area of expertise, namely, loss prevention and process safety, the majority of accidents have well-known causes. Occasionally, an accident occurs because, for example, no one realized that A and B mixed together will react violently under certain conditions. However, such accidents are the exception (and even here it is well established that tests should be carried out). Most accidents are very similar to accidents that have happened before, as the following examples show.

1. To start with a simple mechanical accident, a member of the UK Health and Safety Executive wrote

   It is a chastening thought that, despite all the efforts of inspectors over the years and the accumulated experience of accidents, the belief is still current in some quarters that smooth rotating shafting is not dangerous, and accidents continue to happen on shafting as each generation re-learns the lessons of its predecessors.[1]

2. On plants that regularly use flexible hoses for transferring liquids or gases, people are often injured when removing hoses that still contain some liquid or gas under pressure. They may be injured by the liquid or by sudden movement of the hose. After each incident, it is realized that the pipe to which the hose is connected should have been fitted with a vent valve so that the pressure can be released safely (Figure M10.1). There follows a campaign for fitting vents. After a couple of years, the accident is forgotten and recurs. My book, *Lessons from Disaster—How Organizations Have No Memory and Accidents Recur*[2] describes many similar accidents that are repeated regularly.

3. More serious accidents recur after 10 or more years. People leave, the reasons for the precautions are forgotten, the precautions lapse, and the accident happens again. Chapter 2 of *Lessons from Disaster*[2] describes four fatal accidents that recurred after 10 or more years in the same organization. Here is one of them.

   In 1928 a 36-inch-diameter (91 cm) gas main was being modified, and a number of joints had been broken. The line was isolated from a gas holder by a closed isolation valve that, unknown to those concerned, was leaking. The leaking gas ignited. There was a loud explosion and flames appeared at various joints on the main. One man was killed.

   The source of ignition was a match struck by one of the workers so that he could see what he was doing. However, once an explosive mixture is formed, a source of ignition is always liable to turn up (see Myth 11). The real cause of the explosion was not the match but the leaking valve.

**FIGURE M10.1** Disconnecting hoses has caused many accidents.

144

The following are the conclusions of the original report.

1. Never trust an open gas main that is attached to a system containing gas, and keep all naked lights clear.
2. When working on pipe bridges at night, adequate lighting should be available.
3. Never place absolute reliance on a gas-holder valve, or any other gas valve for that matter. A slip-plate (blind or spade) is easy to insert and absolutely reliable.

In 1967, in another such incident, a large pump was being dismantled for repair. When a fitter removed a cover, hot oil came out and caught fire because the suction valve had been left open. No slip-plates had been fitted because it was not customary practice to fit them. Three men were killed.

The oil was above its auto-ignition temperature. Afterward, the following instructions were issued.

1. Before any equipment is given to maintenance, the equipment must be isolated by slip-plates or physical disconnection unless the job to be done is so quick that fitting slip-plates (or physical disconnection) would take as long *and* be as hazardous as the main job.
2. Valves used to isolate equipment for maintenance, including isolation for slip-plating or physical disconnection, must be locked shut with a padlock and chain.
3. When there is a change of intent, for example, if it is decided to dismantle a pump and not just work on the bearings, the permit to work must be handed back and a new one taken out.

In the period of nearly 40 years that elapsed between these two incidents, the practice of slip-plating had lapsed—no one knew when or why. Perhaps the people who remembered the original incident had left and their successors did not see the need for slip-plating. "It is a lot of extra work," they might have said, "Other companies don't do it."

How can we keep old knowledge alive, especially details of past incidents and the action needed to prevent them from happening again? The following suggestions may help:

1. Remind people of the incidents every few years in safety newsletters and other publications. These should recycle old knowledge as well as describe new.
2. Better still, hold regular discussions of past incidents, their causes, and prevention, as discussed under Myth M5, for employees at all levels.
3. Devise better information storage and retrieval systems so that we can readily locate details of past incidents and recommendations on subjects in which we become interested.[3] Control systems should be able to remind operators of the consequences of approaching hazards, not just sound an alarm, and reports on past accidents and their lessons should be accessible on screen in every control room.[3]
4. On each plant, start a history book containing reports of previous incidents of continuing interest. Do not clutter it up with reports on falls off bicycles and thumbs hit by hammers, but do include reports of interest from other companies.
5. Add a note to each standard and specification explaining *why* it has been adopted.
6. Design plants so that recommendations that are made after an accident can be carried out. (Slip-plating may have lapsed after 1928 because pipe work was not designed with sufficient flexibility for slip-plates to be inserted.)
7. Remember that the first step down the road to the 1967 accident occurred when a manager turned a blind eye to a missing blind.
8. Spend less time reading magazines that tell what is new and more time reading books that tell what is old.

Today, "old" implies outdated; in the past, it implied something of enduring value; it had to be good to have lasted so long.

We cannot make gold from lead; we do not know how. In contrast, we do know how to prevent most accidents that occur. Whether or not we succeed depends on our energy, drive, and initiative and on our ability to access the information that is available.

*References*

1. Watson, G.W. 1983. In *Her Majesty's Inspectors of Factories 1883–1983: Essays to Commemorate 150 Years of Health and Safety Inspection.* London: Her Majesty's Stationery Office.

2. Kletz, T.A. 1993. *Lessons from Disaster—How Organizations Have No Memory and Accidents Recur,* Chap. 10. Rugby, England: Institution of Chemical Engineers; and Houston, TX: Gulf Publishing.
3. Kletz, T.A., et al. 1995. *Computer Control and Human Error in the Process Industries,* Chap 1, Section 5.10. Rugby, England: Institution of Chemical Engineers; and Houston, TX: Gulf Publishing.

## MYTH M11

### To prevent accidents at work, we need to spend more money

Safety by design should always be our aim, but often there is no reasonably practicable way of removing a hazard by a change in design. Instead we have to improve the software, that is, the method of working, the training, instructions, inspections, audits, and so on. We cannot spend our way out of every problem.

Design changes are easy. All that is necessary is money. In contrast, changes to the software are subject to a form of corrosion more rapid than that which affects the steelwork. Procedures can vanish without a trace in a few months once managers lose interest. A continuing effort, what I have called "grey hairs,"[1] is necessary to maintain the software.

Furthermore, design changes do not always cost money. If we make them early in the design stage, we can often *avoid* hazards and do not need to add on safety equipment to keep them under control (see Myth 26).

*Reference*

1. T.A. Kletz. 1982. *Plant/Operations Progr* 5(1):55.

## MYTH M12

### To prevent accidents, we need to persuade reluctant managers and supervisors to take action

Often it is just as necessary and more difficult to persuade foremen and operators to follow the new procedures that managers and supervisors have introduced or to use the new equipment that they have bought. Managers sometimes say, "It must be safe, as we have done it this way for 20 years without an accident" (see Myth 33). Foremen and operators say this more often than managers, and it is harder to convince them that the experience is irrelevant unless an accident in the twenty-first year is acceptable. The manager, among other skills, must be an expert in the art of persuasion.

## MYTH M13

### Policies lead to actions

The managers at the top lay down policy, and the rest of us carry it out. That is the theory, but the practice is often different. We solve our problems, as best we can, subject to various pressures and constraints. Looking back, we see some consistency in our actions. Statements of policy are often a tidying-up operation: putting into statute form what has become the common law of the organization. Actions lead to policies more often than policies lead to actions.

To change the way something is done, say, the way equipment is prepared for maintenance, you could start by persuading the directors to issue a statement of policy. After a serious accident, people are willing to accept direction, and this may be the right solution. At other times, it might be better to persuade individual managers at the lowest level that have the power to change the procedures. When most of them are operating under the new procedures, that is the time to issue a new policy statement.

## MYTH M14

### Treating employees, customers, and suppliers well is good business in the long run

I am not querying the truth of this statement but merely the need to proclaim it as company policy. If a company tells me that they behave decently because it is good business to do so, then they imply that if circumstances change and it is no longer good business, they will stop treating me decently. I approach them cautiously and keep my hand on my wallet.

In contrast, there are other companies, such as Marks and Spencer and ICI in the United Kingdom and Du Pont in the United States, who have established a reputation for dealing fairly with employees, customers, and suppliers at all times, not because it is good business but because it is the right thing to do. The policy is not stated officially but is part of the company culture, rooted in the nature of the organization. Any new manager who tried to change the culture would have a struggle and would probably fail because the culture is deeply entrenched.[1]

We recognize this quality in small businesses as well as large ones. There are one-person repair companies that we trust, and when we deal with them we do not worry about estimates; there are other companies that we treat with caution.

Robert Bolt's play *A Man for All Seasons* is about the life and death of the statesman who opposed Henry VIII's decision to divorce his first wife. The statesman, Sir Thomas More, says, ". . . when statesmen forsake their own private conscience for the sake of their public duties, they lead their country by a short route to chaos."[2] In the same way, when captains of industry forsake fair dealing for a quick profit, they do their shareholders no good in the long run.

*References*

1. Kay, J. 1995. *The Daily Telegraph* (London), 20 Jan.
2. Bolt, R. 1960. *A Man for All Seasons,* p. 22. New York: Random House.

## MYTH M15

### When companies have problems, they should reduce the number of employees

For the last 20 years almost every company in the private sector has repeatedly reduced the number of employees, and we have come to accept this as the obvious reaction in times of trouble. It comes as a surprise to those who were not working in industry in the 1960s to learn that the accepted wisdom at that time, at least in the United Kingdom, was the opposite: if there were problems, we needed more people to help overcome them.

During most of the 1960s, I was working in production, not safety, and whenever we had problems with output, costs, efficiency, quality, or anything else—and we often did—it was considered obvious that we needed extra people to help us solve them. We did not have to ask for them; our bosses thrust them upon us. Temporary buildings to house the larger staff grew on the grass outside the offices. The problem was not reducing staff but recruiting the numbers we wanted. Conventional advertising and university recruiting did not bring in enough potential employees. Groups of recruiters would set up shop in London hotels on evenings and weekends, interviewing anyone who turned up in response to an announcement and offering jobs on the spot to those who were suitable.

Even when we realized that we could reduce numbers, and made a determined attempt to do so, extra people were needed to work out a procedure. In 1966 I was taken off my normal job for a year to head a small team given the task of seeing how we might reduce numbers in the department in which I was then employed.[1] I was given a team of three experienced people, all from the same department, whom I knew well and knew I could work with; there was enough work, I thought, to keep us occupied. I was then told that about six people from the Management Services Department would be joining us. There was no room for more temporary buildings. I went for a walk around the site and found some vacant offices, belonging to another part of the company, about ½ mile away. I was wondering what on earth I was going to give the newcomers to do, but it turned out they had their own ideas. It was a time when work measurement was still popular, and they spent their time calculating the number of units of work actually done and the number of workers theoretically necessary to do it.

Around the end of the 1960s, the nature of the game changed; every effort was made to reduce numbers, and the temporary buildings vanished. The personnel officers who had been kept busy recruiting were now equally busy arranging severance terms and advising on job opportunities elsewhere.

During the severe recessions of the early and late 1980s, numbers were reduced by more than we ever considered possible during our 1966–67 studies. There is no doubt that many companies are managing successfully with far fewer employees than they had then and with far fewer than they believed possible. Nevertheless, perhaps some problems would benefit if more people were assigned to solve them, and perhaps in time it will again become the accepted wisdom to solve at least some problems in this way.[2]

*References*

1. Roeber, J. 1975. *Social Change at Work*. London: Duckworth.
2. Friedlander, R.H., and Perron, M.J. 1995. *Downsizing's effects on safety in the CPI/HPI*. Paper 2c. American Institute of Chemical Engineers Loss Prevention Symposium, August.

# MYTH M16

## Management controls should be sudden and abrupt

I have never seen this view stated, but it is the policy usually followed and is in marked contrast to the philosophy followed in the control of equipment.

If we want to adjust a process plant or an item of equipment such as a furnace, we do not do so suddenly. Instead, we make a small change, check to see that it is producing the desired result, make a further change, and so on. Even if we are sufficiently confident of the result to make a large change in the set point of a controller, it is usually programmed to respond gradually and not to suddenly open or shut the control valve.

In contrast, when we decide on a managerial change, it is usually sudden. We reduce numbers substantially, for example, instead of gradually; we close a factory and discontinue a product rather than experiment with variations. On a national scale, the size and nature of taxes are changed overnight. Those who make the changes are called bold and innovative.

I do not know which procedure is best. I merely question the uncritical assumption that sudden and abrupt change is always right. Perhaps managers, economists, accountants, and politicians should be made aware of proportional control and derivative action, so that they realize there are alternatives to sudden change.[1]

*Reference*

1. Davies, J. 1992. *Chem Engineer* 532/3:6.

# MYTH M17

## A good manager is one who shows the ability to choose between alternatives

If we have to choose between A and B, the best answer is often C. A simple technical problem is presented to illustrate this.

A fuel gas pipeline in a pipe trench went under a road bridge inside a factory. A flanged joint, fitted with a compressed asbestos fiber (CAF) gasket, was located only 2.5 m from the roadway (Figure M17.1). According to the area classification code used by the company, there was a Division (Zone) 2 area for a radius of 3 m

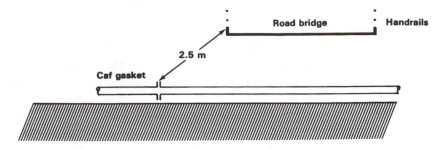

**FIGURE M17.1**   The distance between the gasket and road vehicles was too small.

around a CAF joint, and road vehicles should not be allowed unrestricted access to Division 2 areas. The factory wanted to open the bridge to unrestricted traffic. Two courses of action were suggested.

1. Replace the CAF gasket by a spiral wound one. This would reduce the radius of the Division 2 area to 1 m but would require an inconvenient shutdown of the pipeline, leading to the possibility that the joint, which had never leaked in 10 years, might leak once it had been disturbed.
2. Ignore the code. It is for guidance; it is not law. However, this course of action might give the impression that we disregard our safety codes as soon as they become inconvenient.

   The course of action finally taken was to build a wall along the edge of the bridge in place of the handrail. This increased to more than 3 m the distance that the vapor would have to travel before it reached a source of ignition (Figure M17.2).

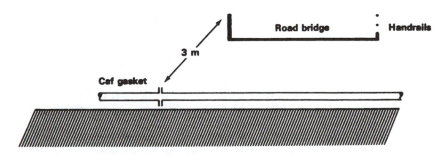

**FIGURE M17.2**   A simple solution to the problem.

Reference 1 describes another problem. During the night, a young, inexperienced supervisor had to decide between aborting a start-up for 24 hours, at a time when product stocks were low, or carrying out an operation that involved some risk, though an experienced foreman had said there would be no risk. There were at least three safe ways of overcoming the problem without aborting the start-up, but under pressure he failed to think of them.

A more fundamental example involves nuclear reactors. In the long run, do we keep nuclear reactors safe by a lot of added-on protective equipment, which may fail or be neglected, particularly in those countries that lack the necessary resources and commitment, or do we do without nuclear-powered electricity? The answer can be "neither": we should build inherently safer nuclear power stations that cannot overheat if the added-on protective systems fail (see Myth 26).

When we are asked to decide between A and B, we seem to focus our attention on the immediate problem to such an extent that we cannot see alternatives. We need to make a conscious effort to tell ourselves that there may be other solutions.

*Reference*

1. Kletz, T.A. 1994. *Learning from Accidents,* 2nd ed., Chap. 3. Oxford, England: Butterworth-Heinemann.

## MYTH M18

### Macho qualities were once so essential for survival that we cannot expect "real men" to act safely

Popular culture has often looked upon cautious behavior as almost cowardly, and on risk taking as more manly. There is the proverb, "Nothing ventured, nothing gained" and the well-known verse by James Graham[1]:

> He either fears his fate too much,
> Or his deserts are small,
> That dares not put it to the touch,
> To gain or lose it all.

(The poet was writing about love, where his philosophy may be justified, but this is not always made clear when he is quoted.)

More prosaically, interviews with managers and other employees in two factories showed that workers who wore all the required protective clothing were considered good and responsible, but also conscientious plodders, slow and reliable, and rather out of touch.[2] There was little or no realization that the time and inconvenience saved by not wearing protective clothing was not worth the risk.

This positive attitude toward risk taking is often said to be bred into us, as our hunter ancestors would not have survived unless some had been willing to risk their lives in search of food. Recent research has cast doubt on this attitude. Jarred Diamond[3,4] measured the number of calories provided by New Guinea hunters and compared it with the number provided by the women (and less macho men) who gather plants and insect larvae. He found that the hunters barely earn their keep and that hunter-gatherers live mainly by gathering. Widows are better off (calorie-wise) than married women; they do not have to keep a man who contributes less than he eats.

Why then is hunting tolerated in New Guinea society? It is not because it supplies necessary protein. Fish, grubs, and small animals such as rats do that more effectively. However, when the hunters make a kill, they "hit the jackpot," and the whole community has a feast. The hunters are heroes and are popular with all the women. Put in the most basic terms, the hunters trade meat for sex and pass on their genes to many children.

There is an analogy here with gambling. We know that unless we are exceptionally lucky, we will lose in the long run, but we still gamble, as the occasional win is so enjoyable; ". . . an action which produces an occasional random success will be repeated far more often than one which carries a predictable outcome."[5]

What is true in New Guinea may not be true in all hunter-gatherer societies or among our ape-like ancestors (though some male chimpanzees give meat preferentially to females in heat[6]); Jarred's work does suggest that there may be an inherited, as well as a cultural, basis for so-called macho attitudes to risk taking. Nevertheless, the New Guinea observations show that this attitude, in at least some jungles, is not necessary for survival. Risk takers are parasites, a burden to others, in the jungle as well as in industry.

*References*

1. Graham, J. (Marquess of Montrose) (1612–1650). *I'll Never Love Thee More.*
2. *Personnel Management,* Feb 1976, p. 25.
3. Diamond, J. 1993. *Natural History* 102(5):24.

4. Diamond, J. 1991. *The Rise and Fall of the Third Chimpanzee,* p. 33–34. London: Random House.
5. Appleby, L. 1994. *A Medical Tour Through the Whole Island of Great Britain,* p. 155. London: Faber and Faber.
6. Stanford, G.B. 1994. *Natural History* 104(1):48.

# MYTH M19

## Stress is a major cause of ill health today

Leonard Sagan[1] argues that the stresses of modern life are small in comparison with those of people living in preindustrial rural societies. We may feel stressed when the boss gives us an urgent deadline, the bills come in, our investments fall, or we read warnings of the effects of all the pollutants in the air we breathe and the food we eat. However, these do not compare with the stresses of burying one's children, of wondering if the harvest will last through the winter, or of seeing one's home washed away by a flood (without any insurance or government aid).

Sagan further claims that the increase in longevity since the Industrial Revolution is not due solely, or even mainly, to improvements in medicine, diet, and sanitation but that the reduction in stress has played a major part. Impoverished people in the West (social class 5) have about twice the death rate of those in social class 1, but this, Sagan suggests, is due to a feeling of helplessness, a feeling that events are not under their control, rather than poor diet or medical treatment. To support his argument, he quotes various studies that show that improvements in diet, medical treatment, and sanitation have not produced the expected decrease in the death rate.

*Reference*

1. Sagan, L.A. 1988. *The Sciences* March/April, p. 21.

# MYTH M20

## The English language has few irregular verbs

Generations of school children in English-speaking countries (and perhaps elsewhere) have struggled with French irregular verbs and consoled themselves with the thought that their own language contains so few. However, it contains more than we at first think. For example,

> I am firm
> You are stubborn
> He is pig headed.

and

> I am a freedom fighter
> You are a guerilla
> He is a terrorist.

So far as safety is concerned, we have

> I show initiative
> You break the rules
> He is trying to wreck the job.

There is a fine line between showing initiative and violating the rules. An action that is praised as initiative when it succeeds may be criticized as a violation if it does not. Sometimes operators are put into a situation of "heads I win, tails you lose." If they follow the rules rigidly, they may be blamed for not using common sense; if they do not follow the rules, and there is an accident, the blame falls on them.

Suppose instructions say that a plant must be shut down when a certain alarm operates. There are a number of false alarms, and many operators get into the habit of ignoring them if everything else seems steady. The supervisor turns a blind eye. One operator, more cautious or obedient than most, shuts the plant down. The alarm turns out to be false, and production is lost. No one can openly criticize the operator, but chance remarks give him (and everyone else) the impression that he was overly hasty. Next time the alarm sounds, it is ignored, and the plant is damaged. Who is to blame?

## MYTH M21

### If something is good, more of it is better

This is the reverse of the next myth (E1) and is certainly not true of chemicals; too much trace metal or vitamins can make us ill. It is also not true of procedures. It is hardly possible to open a safety magazine without seeing a demand for more factory inspectors.

However, if their numbers increase beyond a critical point, effectiveness falls. Managers begin to transfer responsibility to inspectors, thinking "I will not worry about that until the factory inspector complains," or "It must be safe because there is no regulation against it and the factory inspector has not said anything."

What is the critical level? My guess is that it has not been reached in the United Kingdom. However, if there were a substantial increase in the number of inspectors concerned with the chemical industry, we would develop this attitude of overdependence on inspectors.

Similarly, managerial monitoring or auditing is essential; people carrying out routine tasks start cutting corners if there is no one to oversee their work. However, if we watch every move, people say, "Why take care of the details; the boss will pick up any slips I make." Checking everything may not reduce errors; it may increase them.

# 3

# Myths About Toxicology and the Environment

Reading these two books together makes one thing clear: ignoring rigorous science, using sources and opinions selectively, employing illogic and invoking mysticism can permit polemicists to promote wildly different and highly personal viewpoints about the state of the environment.

The above quote is from Jack C. Shultz, referring to his review of two books, one describing a "catastrophic" problem in US forests, the other claiming that the environment has never been better (see the August 1995 issue of *Natural History,* p. 64).

Much discussion on toxicology and the environment is in the language of polemic rather than science. According to Warren Newman, "We are living in a period where in the West environmentalism has become a religion and progress is to be sacrificed in its name."[1] Anyone querying what some regard as self-evident is liable to be denounced as a heretic.

With the diffidence of a 19th-century agnostic daring to doubt religious dogma, I set down in the following pages some doubts about the accuracy of several statements often made about pollution and its effects. I hope I shall not be burned at the stake or condemned to live near the seashore, where I may be drowned by rising water levels, an appropriate punishment for anyone who dares to doubt the extreme estimates of the effects of global warming.

I have not done original work in this field and have had to rely on the views of many others. I have preferred to quote the opinions of people who have spent many years studying the environment and thus have some authority for their views.

In some cases the people I quote are employed by the industries concerned and may therefore be considered biased (see Myth E1). If we want to consult experts on widgets, we often find that they all work for widget manufacturers. Nevertheless, my sources have usually had to justify their published views by the usual method of peer review. Doll and Peto, in the paper quoted in Myth E5, take 200 pages to justify the conclusions that I have summarized.

In contrast, many of the statements made by "green" campaigners are based on anecdotal evidence or a single piece of work that has not been confirmed by later studies.[2]

The myths on technology and management in the first two parts of this book are accepted as correct by many scientists and engineers. In contrast, few scientists or engineers believe the myths in part 3. There is a large gap between the scientific consensus and the views presented by the media, though these views are accepted by many members of the general public, including politicians. A cynic might say that their merit is not accuracy but the ability to attract readers and voters. Certainty sells, and while many environmentalists seem to have no doubts (at least in public) about the accuracy of their views, the views of scientists are, as always, full of qualifications. It is easy to say, on the basis of a few observations, that something is harmful to the environment, and it is impossible to prove with complete certainty that it is not.

We should not forget that people in the West have never been healthier than they are today nor lived longer, that town air is cleaner than at any time since the industrial revolution, that pesticides promote good health, not disease, and that plastics, man-made fibers, and other synthetic products improve the quality of life in innumerable ways. This is no excuse for not continuing to improve the environment, but at the same time, we should weigh in the balance the cost of improvements, the benefits they will bring, and the other uses to which tax money might be put.

I am not, of course, opposed to attempts to improve the environment, only to the excesses of the environmental movement. Most scientists and engineers welcomed the campaigns for cleaner air, following the London smog of 1952, and the opposition to above-ground nuclear tests in the 1950s and early 1960s. These campaigns against real hazards now seem very low key compared with the almost hysterical opposition to selected small or negligible risks that we see today. There is still no shortage of hazards that kill many people and reduce the quality of life (road accidents, drugs, smoking, alcohol, poor diet, accidents in the home), but many environmentalists seem uninterested in them.

The turning point seems to have been the publication in 1962 of Rachel Carson's *Silent Spring*. This mixture of science and fiction emphasized the hazards of DDT, a pesticide that may have saved more lives than any other man-made chemical (though chlorine may have saved more). The results of banning DDT can be seen by looking at the cases of malaria in Ceylon[3]:

| Cases | Year | Comment |
|---|---|---|
| 2,800,000 | 1948 | No DDT |
| 31 | 1962 | Large-scale DDT program |
| 17 | 1963 | Large-scale DDT program |
| 150 | 1964 | Spraying stopped |
| 308 | 1965 | |
| 499 | 1966 | |
| 3466 | 1967 | |
| 1,000,000 | 1968 | |
| 2,500,000 | 1969 | |

When science and technology have been applied, they have increased the lifespan and quality of life beyond the dreams of earlier generations, but it has not been all gain. A small fraction of the gain has been offset by events such as Bhopal (where more than 2000 people were killed by a toxic chemical leak; see Myth 20) and by the effects of some chemicals such as thalidomide, asbestos, and lead. We must try to foresee and prevent these adverse effects, but it is illogical to assume that we can or should go back to the ways of pretechnology.

*References*

1. Newman, W. 1991. *Atom* 415:14.
2. For examples, see Whelan, E.M. 1993. *Toxic Terror.* Buffalo, NY: Prometheus Books.
3. Whelan, E.M. 1993. *Toxic Terror,* p. 101. Buffalo, NY: Prometheus Books.

## MYTH E1

### If something is harmful, less of it is proportionately less harmful

We know that large doses of radioactivity and toxic chemicals are harmful, and so we assume that a thousandth part of a harmful dose may also cause harm, though it is one thousand times less likely to do so.

However, suppose we make our starting point small doses instead of large ones. Small doses of ultraviolet light, vitamins, salt, arsenic, and certain heavy metals are harmless, indeed good for us, but large doses are harmful (except vitamin C). Most drugs are harmful if we exceed the stated dose. By this reasoning, perhaps radioactivity and toxic chemicals are harmless, perhaps even beneficial, in small doses? Perhaps the body can cope with small doses that become harmful only when the body's defense mechanisms are overwhelmed. Since we are all exposed, all the time, to small doses of natural radioactivity, perhaps the body has developed a mechanism for coping with its effects.

John Fremlin has surveyed the evidence that small doses of radioactivity, especially beta and gamma radiation, are beneficial, and while the case is not proven, there is some evidence to support this view.[1,2] He discusses possible mechanisms and says that no substance has ever been shown to have a proportionately less harmful effect at very low doses.[1]

Large doses of vitamin A are poisonous. Arctic explorers died as a result of eating the livers of animals such as polar bears that are at the top of the food chain, so that the vitamin accumulates in the liver.[3] Suppose we had discovered the toxic effects first, before we discovered the beneficial effects. Would we have tried to limit the consumption of vitamin A?

It is usually impossible to prove that small concentrations of toxins are harmful, as any effects would be too small to be seen above the background noise. Thus the belief that they are harmful is not a theory (something that can be proved or disproved), but an act of faith or, at best, a reasonable but conservative assumption for use in fixing tolerable levels of exposure. Two case histories are presented below by way of illustration.

1.  During the 1980s the plant growth regulator Alar[4-6] was tested on rodents. As usual during such tests, the doses used were very high,

the maximum tolerable. During tests to assess noncarcinogenic tox-icity, the usual procedure is to measure the amount that produces no effect in the test animal and then apply a large safety factor, typically a hundred. Only when testing for cancer is it usual to feed the maximum possible dose to the test animal. It is not clear why this entirely different procedure was adopted.[7] Some of the rodents de-veloped cancer. As about half the chemicals, natural and synthetic, tested in this way produce cancers, it is possible that the cancers are due to gross overfeeding rather than the toxic action of the chemical. Sections of the public became concerned that consumption of apples treated with Alar could cause cancer. Apple products were destroyed, and schools banned the fruit. According to a press report, a mother, having heard that apples cause cancer, telephoned the police and got them to stop a school bus and remove an apple from her daughter's lunch box![6]

The public overlooked, or was not told, that it would be necessary for a child to consume 18 $m^3$ of apple juice (the contents of two tank trucks) per day for a lifetime in order to receive a dose of Alar similar to that which causes cancer in rodents. Nevertheless, a campaign against Alar by an environmental group, backed by a television reporter and two U.S. senators, was successful. Alar was withdrawn, first in the United States and then from the export trade.

Part of the environmental group's case was that the experts in support of Alar worked for or were paid by the manufacturers, while the experts from the environmental groups had no bias. On the other hand, the reports criticizing Alar never underwent rigorous peer review. The opponents of Alar, headed by a professional publicist, were far more skilled at presentation than the defenders.

2. The drug Opren[8] is most effective at relieving the pain of arthritis. However, it produced side effects in a few people: it made them very sensitive to sunburn, and a few elderly people died. After a media scare, the drug was taken off the market, to the dismay of many sufferers from arthritis, who had to make do with less effective pain killers.

The chance of being killed by taking Opren is 1 in 25,000, so the risk was never detected during testing. The chance of dying as the result of surgery for arthritis is 1 in 60, which is 400 times greater, yet surgery is accepted, but Opren is banned. The newspapers that publicized and exaggerated the risk of Opren have been responsible for a great deal of suffering (see Myth E7).

Drugs that have been used for many years would ironically not be approved if they were introduced under the testing standards required today, for example, aspirin, penicillin, and salvarsan (the original anti-syphilis drug which contains arsenic and is kept in reserve in case the bacteria develop resistance to penicillin).[8]

*References*

1. Fremlin, J.H. 1989. *Atom* 390:4.
2. Fremlin, J.H. 1989. *Power Production—What Are the Risks?*, 2nd ed., pp. 57–68. Bristol, England: Hilger.
3. Howell, M., and Ford, P. 1985. *The Ghost Disease,* Chap. 12. London: Penguin Books.
4. Foulkes, D.M. 1992. In *Risk Management of Chemicals,* ed. M.L. Richardson, p. 307. London: Royal Society of Chemistry.
5. Jasanoff, F. 1992. In *Risk Management of Chemicals,* ed. M.L. Richardson, p. 307. London: Royal Society of Chemistry.
6. Whelan, E.M. 1993. *Toxic Terror,* p. 18. Buffalo, NY: Prometheus Books.
7. Whelan, E.M. 1993. *Toxic Terror*, p. 62. Buffalo, NY: Prometheus Books.
8. Emsley, J. 1994. *The Consumer's Good Chemical Guide,* pp. 132, 139, 142. New York and Oxford, England: Spektrum.

# MYTH E2

## The most hazardous substances are those labeled "dangerous"

Only a small fraction of the substances that exist naturally or have been synthesized have ever been tested. Many of these substances may be, and in fact probably are, as toxic as those that have been tested. If a substance has been tested and is labeled accordingly, then we know the hazards and know what precautions to take. If a substance has not been tested, we cannot be warned.[1]

In many countries, regulations require substances that are offered for sale to be tested, but these regulations do not apply to naturally occurring substances that have long been available, such as most of the constituents of the food we eat.

Workers in the nuclear power industry are usually considered to be exposed to hazards, and extensive precautions are taken. Nevertheless the workers in a Belgian nuclear power plant showed more changes in their blood cell chromosomes than office workers. The workers in a fossil fuel power plant showed even more changes in their chromosomes than the nuclear plant workers. This is not really surprising, as many products of combustion are carcinogenic.[2]

Common sense tells us that substances that are not labeled "dangerous" (or generally recognized as dangerous) are safe. Common sense can be wrong.

### References

1. Breyer, S. 1993. *Breaking the Vicious Circle,* p. 56. Cambridge, MA: Harvard University Press.
2. Evans, H.J. 1994. In *The Candle Revisited,* eds. P. Day and R. Catlow, p. 127. Oxford, England: University Press.

## MYTH E3

### We should look for the effects of substances that have not yet been studied

If we do not know the effects of a substance that is offered for sale or that is in regular use, should we find out? After an accident, we look for causes, so why do we not look for the causes of toxic effects instead of starting with substances and looking for their effects? For example, we looked for the effects of lead in gasoline, found that it caused low IQs in children, and decided to remove it, or at least reduce the amount used. Suppose we had started instead at the other end and asked what is the cause of low IQs in children. One cause is lead in gasoline, but there are others: lead in paint and lead water pipes, neurotic symptoms in mothers. Perhaps the money spent in removing lead from gasoline would have had more effect if it had been used, in whole or in part, to remove some of the other causes. I am not saying it should, only that the question should be asked.[1]

Broadbent[1] suggests that if we spent more effort on identifying and publicizing the major causes of cancer, perhaps we would save more lives than we will ever save by stopping minute discharges of radiation or carrying out extensive tests on drugs and pesticides.

Baklien writes[2]

> It is now estimated that close to one million scientists, lawyers and bureaucrats are occupationally concerned with drugs and pesticides as potential carcinogens, in spite of the fact that with minor exceptions (mostly anti-cancer drugs) such chemicals have not been known to cause cancer in man.

Perhaps we might by now have a cure for cancer, had we put half of these people onto cancer research.

*References*

1. Broadbent, D.E. 1985. In *Man-Made Hazards to Man,* ed. D.E. Cooper, p. 54. Oxford, England: Clarendon Press.
2. Baklien, A. 1981. *Search* 12(1-2):30.

# MYTH E4

## Natural materials are safe

There is widespread belief among the general public that naturally occurring compounds are harmless or at least less harmful than man-made pesticides, antioxidants, and other additives. The work of Bruce Ames and co-workers[1-4] has shown that this is not true. Most naturally occurring compounds have never been tested.

Of those that have been tested, about half are carcinogenic, the same proportion as for synthetic chemicals. However, our diet is made up mainly of natural compounds and contains only traces of synthetic additives. The average U.S. diet contains about 0.1 mg per day of synthetic pesticides but 1500 mg per day, 15,000 times as much, of known natural pesticides, about half of which are carcinogens.

Common foods that contain known poisons are alcohol, apples, bacon, bananas, basil, beer, broccoli, Brussels sprouts, cabbage, cantaloupe, carrots, cauliflower, celery, chocolate, cinnamon, cloves, cocoa, to begin alphabetically. None of these would be approved for human consumption if they were tested in the same way as pesticides and other additives. However, there is no proof that the amounts in a normal diet will cause harm. As stated in Myth E1, the body may be able to cope with small amounts of harmful substances, and they may produce effects only when the body's defenses are overwhelmed.

Has the body, during the course of evolution, developed resistance to natural poisons? In a few cases, such as radioactivity (see Myth E1) it may have done so, but our diet has changed drastically during the last few thousand years, and our bodies have had too little time to develop resistance to the new foods that we now eat.

Why do plants contain so many natural toxins? It is their main defense against predators. They cannot pull up their roots and run away. All they can do is poison (or prick) their enemies or develop unpleasant tastes.

What are the benefits of the various additives used during food growth and afterward? Natural pests (insects, diseases, and weeds) are said to cause the loss, before harvesting, of about a third of the food that is grown and another fifth afterward.[5] Housewives might be less willing to buy food that is labeled "contains no additives" if instead the label said, "contains nothing to prevent it spoiling" or if all the natural toxins were listed. People who avoid fruit and vegetables because they may be contaminated by traces of pesticides increase their chances of developing cancer, as many fruits and vegetables contain anti-carcinogenic compounds. It is paradoxical that the people who object to food additives often prefer processed foods instead of natural ones, for example, they may prefer low-fat or polyunsaturated spreads instead of butter, and texturized vegetable protein instead of meat.

*References*

1. Ames, B.N. 1983. *Science* 221:1256.
2. Ames, B.N. 1989. *Chemtech* Oct, p. 590.
3. Ames, B.N., and Swirsky, L. 1990. *Science* 249:970.
4. Gold, L.S., et al. 1992. *Science* 258:261.
5. Pimenlel, D. 1991. *Chem Britain* 27(7):646.

## MYTH E5

### The causes of cancer are environmental

If we use the word "environmental" in its original sense, to mean that the causes lie outside the human body, then it is true that the causes of cancer are environmental. However, if we use "environmental" to mean that cancer is caused by pollution of our surroundings, then it is not true or, to be precise, only about 2% true.

In their exhaustive study of the causes of cancer deaths, Doll and Peto[1] suggest that about a third of the deaths are due to smoking and another third could be prevented by changes in diet. Possibly 10% are due to infection and 7% to sexual habits. About 4% have industrial causes, and 2% are due to pollution. Other causes (alcohol, medicines, ultraviolet light, and radioactivity, mainly natural) account for the remaining 10%.

There is, of course, a good deal of uncertainty in these figures. The number of cancer deaths due to occupational causes could be as high as 8% or as low as 2%; the number due to pollution could be as high as 5% or less than 1%. The number due to food additives is less than 1% and probably negative, that is, food additives prevent cancer by preventing the deterioration of food.

The occupational deaths are concentrated among comparatively few people who are exposed to relatively high risk; about three-quarters of the deaths are due to lung cancers, mostly from asbestos. Their prevention therefore justifies higher priority than the absolute number might suggest.

To convert these percentages to actual numbers, remember that there are about 400,000 deaths per year from cancer in the United States,[1] about 140,000 in the United Kingdom.[2]

*Reference*

1. Doll, R., and Peto, R. 1981. *J. National Cancer Institute* 66(6):1194.
2. Office of Population Censuses and Surveys. Annual reports, *OPC's Monitor,* ref. DH2. London: Her Majesty's Stationary Office.

## MYTH E6

### The public spends its money unwisely

Politicians and other public figures often tell us that the public, especially those of low income, wastes its money on alcohol, tobacco, and junk food instead of buying fruit, vegetables, and healthy whole-wheat bread. Government money is spent, with little effect, to advertise and advise people as to a healthy diet. The accusation may be true, but the extent to which people misspend their own money is nothing compared with the extent to which governments misspend the public's money.

In the developed world, doctors spend thousands or tens of thousands of dollars to save one life, road engineers spend hundreds of thousands, the chemical industry millions, and the nuclear industry tens of millions (more according to some estimates) (see Table E6.1). In developing countries, lives can be saved for quite trivial sums. Governments directly authorize or influence much of this expenditure (to an extent that varies from country to country) without consideration for the effectiveness of alternative ways of spending resources. No other commodity or service shows such a wide variation in cost.

**TABLE E6**   Estimates of the money spent to save one life, 1985 prices in the United Kingdom (in U.S. dollars)

| What Money is Spent On | Amount of Money Spent (K = thousand, M = million) |
| --- | --- |
| *Health* | |
| Increasing tax on cigarettes | Negative |
| Antismoking campaigns | Small |
| Cervical cancer screening | 10K |
| Artificial kidney | 60K |
| Intensive care | 30K |
| Liver transplant | 150K |
| *Road travel* | |
| Various safety measures | 30K – 12M |
| Safety measures implemented | Up to 1.5M |
| *Industry* | |
| Agriculture (employees) | 15K |
| Rollover protection for tractors | 600K |
| Steel handling (employees) | 1.5M |
| Pharmaceuticals (employees) | 30M |
| Pharmaceuticals (public) | 75K |
| Chemical industry (employees) (typical figure) | 6M |
| Nuclear industry (employees and public) | 25–45M |
| *Social policy* | |
| Smoke alarms | 750K |
| Preventing collapse of high-rise apartments | 150M |
| Giving members of social class 5 a social class 2 income (family of 4) | 1.5M |
| Third world starvation relief | 15K |
| Immunization (Indonesia) | 150 |

Notes

All figures are taken from reference 1, are corrected to 1985 prices, and refer to the United Kingdom; U.S. figures are somewhat higher. All figures are approximate, and some may now be outdated due to changes in technology.

A 10% increase in the tax on tobacco decreases smoking by about 5%, so there is a net increase in revenue.

If we spend $10M on antismoking campaigns and as a result 1000 people (less than 1 smoker in 10,000) give up smoking, the cost of saving a life will be about $10K.

The death rate (for almost all ages and causes) of members of social class 5 (unskilled occupations) is about 1.8 times that of members of social classes 1 (professional occupations) and 2 (managerial occupations). It can be argued that in the long run, a rise in income to the social class 2 level will produce a social class 2 lifestyle.

Most of the values in Table E6.1 are unknown to those who authorize the expenditure, as they rarely divide it by the number of lives saved. But the figures are available for them to do so.

There are pressures on governments to be considered. In a democracy the government should meet its public's needs. If the public prefers to see its money spent on making small risks even smaller, the government, as a good servant of the public, should provide funds for that cause. However, governments often assume that noisy pressure groups and people who have axes to grind represent public opinion.

The public does not always allocate resources wisely. Governments rarely do.

*Reference*

1. Kletz, T.A. 1988. In *Engineering Risk and Hazard Assessment,* eds., A. Kandel and E. Avni. Boca Raton, FL: CRC Press.

## MYTH E7

### Scientists should be held responsible for the results of their actions

All of us should be held responsible for our actions, morally if not legally (while recognizing that we will inevitably make slips and have lapses of attention from time to time, and are not always properly trained; see Myth 6). However, it is erroneous to single out scientists for special mention. Do their errors and mistakes produce more havoc than those of other sectors?

The destruction of ozone by chlorofluorocarbons (CFCs) and bromofluorocarbons (BFCs or halons) is an example of the serious unforeseen results of what was at first regarded as a major scientific breakthrough. CFCs were at first regarded as wonder chemicals, efficient, nonflammable, nontoxic refrigerants, solvents, and aerosol propellants, while BFCs were excellent fire-fighting agents. (In 1937, Thomas Midgley demonstrated the safety and effectiveness of CFCs by filling his lungs with the vapor and then exhaling, extinguishing a candle.[1]) Only gradually did we become aware of their disadvantages. There have been other, less serious, changes that had unforeseen side effects, and there is no doubt that scientists and engineers should carry out systematic searches for the possible effects of change using techniques such as hazard and operability studies[2] (see Myth 25).

Everyone who advocates change should do the same (though they will need different techniques). Everyone should accept responsibility for his or her actions. Newspaper reporters and television producers may overreport on a drug and secure its withdrawal because it produces unpleasant side effects in a few people. Are the reporters and producers not responsible (morally) for the suffering of those who could have benefited from the drug? (see Myth E1). Lobby groups and the media exaggerate the possible effects of chemicals, small doses of radioactivity, or radiation from electric cables to such an extent that vast sums are spent on investigations, money that could produce far more benefit if spent in other ways. Of course, they all mean well, but their actions can produce a net increase in suffering. Are they held responsible (or do they hold themselves responsible) for the results of their actions? Do they weigh in the balance the possibility that they might be wrong and that resources will be wasted? An example is the furor over Love Canal, an unfinished and abandoned canal used as a dump for toxic chemicals and later developed for housing. In fact, there was no more illness in the area than would normally be expected, as the *New York Times* ultimately admitted (". . . it may well turn out that the public suffered less from the chemicals there than from the hysteria generated by flimsy research irresponsibly handled").[3] Who was responsible for the hysteria that led people to leave their homes: those who deposited the chemicals, those who allowed homes to be built on top of the dump (despite agreeing not to do so), or those who spread unproven stories?

Michael Frayn[4] has described three stages in the history of ideas. They also apply to inventions and new substances.

1. When they are new, they are interesting, challenging, and about to open up a wonderful new age.
2. When they are mature, they are inhuman, soul-destroying, contemptible, and ridiculous.
3. In their old age, they are regarded with understanding and affection and are studied and collected.

### References

1. Hendershot, D.C. 1995. *Process Safety Progr* 14(1):52.
2. Kletz, T.A. 1992. *Hazop and Hazan—Identifying and Assessing Process Industry Hazards,* 3rd ed., Chap. 2. Rugby, England: Institution of Chemical Engineers.

3. Whelan, E.M. 1993. 20 June 1981, *New York Times,* quote. In *Toxic Terror,* Chap. 3. Buffalo, NY: Prometheus Press.
4. Frayn, M. 1990. *The Original Michael Frayn,* p. 197. London: Penguin Books.

# MYTH E8

## Plastics should be made biodegradable, or we will soon run out of disposal area in the ground

Do we really want our gutters, plastic window frames, and plastic water pipes to biodegrade (that is, rot)? When plastics biodegrade, they form carbon dioxide and increase the greenhouse effect (see Myths E12 and E13). Is it not better to keep the carbon locked up in the plastic?

If trash is incinerated, plastics usually generate less carbon dioxide than paper or wood; plastic bags form less carbon dioxide than paper ones (and have the additional advantages that they keep their contents dry and do not lose their strength when wet). The main advantage of biodegradable plastics is that plastic litter will disappear more quickly than that of nonbiodegradable plastics.

Some countries, such as Holland and Israel, are short of disposal space in the ground. In Israel a large hill on the edge of Tel Aviv is a mountain of waste. Many countries, however, are making new holes for disposal, as the result of quarrying and mineral extraction, faster than they are filling them.

Whether or not a country is short of disposal space in the ground, making plastics biodegradable will have little effect on the total amount of waste produced. In the European Community, plastics amount to only 7% of domestic waste, and domestic waste to only 4% of the total waste produced, most of it being industrial.[1]

Of course, plastic should be recycled when possible, and many goods could be packaged in less plastic. I re-use old plastic carriers when I go shopping, something I could not do with paper ones when it is raining.

*Reference*

1. *The Roundel* 1991. 69(6):151.

# MYTH E9

## We should recycle whenever possible

We should recycle when it is economical to do so, and perhaps be prepared to pay a premium to avoid using up limited resources.

However, we should also know the mechanics of recycling. A letter in the *Wall Street Journal*[1] claimed that making a ton of paper from recycled newsprint uses nearly a ton of oil (7 barrels) more than making it from virgin pulp. Much of the extra oil is used in transporting the waste paper to the nearest paper mill. All the expended oil will end up as carbon dioxide in the atmosphere.

Some years ago I listened to a newsprint manufacturer explain how he had made the use of recycled paper economical by persuading volunteers to collect it for nothing.[2] He showed commendable enterprise, but how many other ailing industries would survive if volunteers would work for nothing, and would the volunteers do more good if they did volunteer work for other ailing industries?

*References*

1. Kirby, R. 1993. *Wall Street Journal.* 7 July.
2. de Jong, R.L. 1984. *CHEMRAWN III World Conference on Resource, Material, Conversion.* Paper 3.V.3. Amsterdam, June.

## MYTH E10

### The polluter should pay the costs of pollution

To say "The polluter should collect the costs of pollution" is fair; we should know the true costs of each product and activity, including the costs of removing pollutants or repairing the damage they cause. However, the polluter is just a middleman; he may appear to pay the costs of pollution but he actually passes them on to the rest of us in the price of the goods and services he provides. We consumers pay in the end whether we pay the costs directly, pay the polluter when we buy his goods or services, or pay through our taxes.

Every time we ask for more money to be spent, by government or industry, on preventing pollution, increasing safety, or carrying out some other worthwhile task, we should ask what the costs will be to us personally. The media often report that the government will have to find the money for this and that. They do not say to the reader or listener, "you will have to pay." Often the cost is small and well worthwhile. Sometimes it is large but still worthwhile, but not always.

At present, some pollution costs are internalized (that is, added to the price of the goods or services), and others are not. For example, the cost of nuclear electricity includes the cost of removing almost all traces of radioactivity from the air and water. The costs of electricity made from coal do not include the cost of repairing the damage caused by sulfur dioxide or (as a rule) the loss of value of the land surrounding a colliery waste tip. To take a trivial example, by using disposable plates, fast food outlets save the cost of crockery and dishwashing and pass the savings onto their customers. However, everyone has to pay for the cost of picking up the litter from the street.

## MYTH E11

**Nothing should stand in the way of an opportunity to improve the environment**

I have stated this myth in a rather extreme form. Most environmentalists do accept that money is limited and that we cannot do everything at once. However, their commendable enthusiasm (and visible campaigning) to prevent pollution has resulted in some hasty and ill-considered measures that have had adverse effects on safety.[1]

- When the effects of CFCs (mentioned in Myth E7) on the ozone layer were realized, aerosol manufacturers started to use butane (or propane/butane mixtures) instead. The result was a series of fires and explosions. Did the changeover have to be made so quickly and with so little consideration of the hazards of handling hydrocarbons? The reports on some of the fires that have occurred say that the manufacturers did not understand the hazards and that elementary safety precautions were lacking. One UK company was prosecuted for failing to train employees in the hazards of butane, in fire evacuation procedures, and in emergency shutdown procedures. These actions were, of course, not necessary or were less necessary when CFCs were used. After this fire, factory inspectors visited other aerosol factories and found much that could be improved. Manufacturers agreed to modify filling machines so that they were suitable for handling butane. This, apparently, had not been considered before.

  There has also been a move to use light hydrocarbons or ammonia instead of CFCs as refrigerants by people who do not know (or perhaps knew and chose to ignore the facts) that both are flammable and that ammonia is also toxic. A television report on the 1994 Winter Olympics said that the bobsled runs used "safe" ammonia gas as refrigerant.[2]

- A number of compressor houses and other buildings have been destroyed or seriously damaged, and the occupants killed, when leaks of flammable gas or vapor have exploded. A building can be destroyed by the explosion of a few tens of kilograms of flammable gas used indoors, but out-of-doors, several tonnes or tens of tonnes are needed. For this reason, during the 1960s and 1970s most new compressor houses and many other buildings in which flammable materials are handled were built without walls, so that natural ventilation could disperse any leaks that occurred. The walls of many existing buildings were pulled down.

  In recent years many walled-in buildings have again been built in order to meet new noise regulations. The buildings are usually provided with forced ventilation, but this is much less effective than natural ventilation and is usually designed for the comfort of the operators and is not powerful enough to disperse leaks rapidly.

- During the last 20 years there has been increasing pressure to collect the discharges from tank vents, gasoline filling, etc., instead of discharging them to the atmosphere, particularly in areas subject to photochemical smog. A 1976 report said that when gasoline recovery systems were installed in the San Diego area, there were over 20 fires in 4 months. In time, the problems were overcome, but it seems that the recovery systems were introduced too quickly and without sufficient testing.

- Shortening of pipe runs to avoid heat losses and save fuel can result in congested plants; if a fire occurs, more damage is done, particularly if equipment is stacked above pipe runs. There is no net gain if we save fuel bit-by-bit and then waste the savings in a big display of fireworks.

- Collecting dust to reduce pollution and collecting them dry to avoid water pollution has led to dust explosions.

*References*

1. Kletz, T.A. 1993. *Process Safety Progr* 12(3):147.
2. Hendershot, D.C. 1995. *Process Safety Progr* 14(1):52.

## MYTH E12

### The atmosphere acts like the glass in a greenhouse and keeps the earth warm

About 1830 the French mathematician Fourier argued that the atmosphere acts like the glass in a greenhouse by letting through the short-wavelength radiation from the sun but absorbing the longer wavelength radiation from the earth. (The cooler an object, the longer the wavelength of its radiation, so the radiation from the earth has a longer wavelength than that from the sun.) Fourier was right about the earth but wrong about greenhouses. The glass helps a greenhouse stay warm by preventing the warm air inside, heated by the sun, from mixing with the cooler air outside, not by trapping radiant heat.[1]

The term "greenhouse effect" is thus a misnomer, but it is now so established that it has come to stay.. The phrase "greenhouse gases" is used to describe those gases, mainly water vapor, carbon dioxide, and methane, that absorb the radiation from the earth. As the amount of carbon dioxide in the atmosphere increases, as the result of burning forests and fossil fuel, so it is argued, the earth will get warmer (see next myth).

*Reference*

1. Moomaw, W.R. 1989. *Orion Nature Quarterly* 8(1):5.

## MYTH E13

### Fuel economy can prevent the catastrophic results of the greenhouse effect

This statement combines two half-truths. First, it is not certain that the results of the greenhouse effect will be catastrophic. There is no doubt that the amount of carbon dioxide in the atmosphere has risen, from an estimated 250 ppm before the industrial revolution to about 350 ppm today.[1] About half this rise has occurred since 1958 as the result of increased burning of forests and fossil fuels, and the carbon dioxide level will continue to rise if we go on burning fuel at the present rate. There is, however, no agreement on what the results will be. Many writers have claimed that global temperatures will rise with devastating effects on climate, that sea levels will also rise, and that much land will be flooded. There may be positive feedback, that is, a warmer climate might cause more water to evaporate and more methane to be released into the atmosphere from the frozen soils of the Arctic regions, both occurrences adding to the greenhouse effect. (There is about 8 times more carbon in the soil, plants, and the sea than in the atmosphere.)

Other writers,[2] less numerous, suggest that there may be negative feedback: increased evaporation may increase cloud cover and restrict the rate of temperature rise and may also increase absorption of carbon dioxide by the oceans. They also remind us of two facts that are often overlooked.

1. If all the Arctic ice were to melt, it would have little or no effect on sea level, as the ice is already floating, most of it below sea level. The small amount of ice that is above sea level will be absorbed by the contraction in volume as the ice melts, since the weight of the ice equals the weight of water displaced. (Antarctic ice would, of course, raise sea level if it melted.)
2. Carbon dioxide absorbs infrared radiation at two wavelengths only, 4.3 and 25 microns. It is transparent to other wavelengths. However much we put into the atmosphere, its absorbing power is limited. Even if its concentration in the atmosphere were to double, the temperature of the earth would rise by only about 1°C, not the 5°C often quoted. Water vapor and methane are far more efficient as greenhouse gases. We cannot control the amount of water vapor at all and have little control over methane, as most of it is generated by rotting organic matter.[3]

Because no one knows for certain what will happen, it seems wise to limit the rise in atmospheric carbon dioxide. Governments have agreed but have done little. What can they do?

Economy and renewable resources could play a significant part. Fuel is used extravagantly in the developed countries. However, fuel economy and renewable resources can never achieve the desired reductions in carbon dioxide discharges. In addition, developing countries want a better standard of living and that means more demand for fuel, including electricity. The developed countries, having achieved a high standard of living, can hardly deny fuel to the rest of the world. The only way we can maintain, let alone increase, electricity production without a rise in carbon dioxide is by investing heavily in nuclear power, as was done in France.

To quote Warren Newman,[5] "If you accept that there is a moral case for recognizing the overriding need of poor nations to use more fossil fuels then western nations must go the French route and develop advanced non-polluting sources of electricity."

There is concern that developing nations may not have the resources, skills, and commitment needed to maintain the complex added-on protective systems that form part of typical nuclear power stations. However, there are inherently safer designs of nuclear power plants that are not dependent on complex added-on safety systems, which may fail or be neglected. They include the high-temperature gas reactor, in which overheating is prevented by a combination of high surface-to-volume ratio and high-temperature resistance, and the Swedish PIUS (process inherent ultimate safety) reactor, which is immersed in boric acid solution. If the cooling pumps fail, the solution is drawn into the core by convection. The boron absorbs neutrons and stops the chain reaction, while the water removes the residual heat. These designs are not yet commercially available, but their development for use in developing countries should be encouraged.[4]

In the United Kingdom the most that can be saved by fuel conservation is probably no more than 10% of present consumption, and the most savings that can be obtained from renewables is probably about the same. (In 1994, renewables supplied 2%.[5,6]) Other developed countries will have similar statistics, except for those that can produce hydroelectricity. Nuclear power can reduce carbon dioxide emissions at much lower cost than many of the energy saving measures that have been proposed. In practice, we may need both. Conservation and renewables alone cannot prevent a further rise in atmospheric carbon dioxide, let alone compensate for the growing demand for electricity in the developing countries.

In the very long term, say, thousands of years, increased levels of greenhouse gases in the atmosphere may be the only way to prevent a return of the ice age. If we build up a large reservoir of carbon, it could then be burned. We can stockpile carbon by preserving and extending the world's forests. Other reasons for preserving forests are the effects on climate, prevention of soil erosion, preservation of habitats and the species that live in them, as well as long-term foresight. Boreal forests are at risk as well as tropical ones.

It is often alleged that acid rain is killing Europe's forests, but it is also claimed that, as a result of the increase in carbon dioxide in the atmosphere and fertilization by acid rain, the forests are growing faster than ever. If this is true, it could explain the discrepancy between the amount of carbon dioxide formed by combustion and the rise in carbon dioxide in the atmosphere. After allowing for absorption by the oceans, about a quarter of the carbon dioxide formed is unaccounted for.[7]

*References*

1. Wolf, E.C. 1989. *Orion Nature Quarterly* 8(1):37.
2. Idso, S.B. 1992. *Speculations Sci Technol* 15(2):123.
3. Emsley, J. 1994. *The Consumer's Good Chemical Guide,* Chap. 9. New York and Oxford, England: Spektrum.
4. Kletz, T.A. 1991 *Plant Design for Safety—A User-Friendly Approach,* Chap. 11. Washington, DC: Taylor & Francis.
5. Newman, A. 1991. *Atom* 415:14.
6. See papers by Fells, I., and Butler, G. 1994. In *Engineering: A Key Aspect of the UK Nuclear Policy Review,* eds. J. Bindon and S. Butcher. London: Institution of Electrical Engineers. A paper by P. Green, opposing nuclear power, was, unlike those by Fells and Butler, entirely qualitative.
7. Emsley, J. 1995. *Chem Britain* 31(7):562.

## MYTH E14

### Early peoples lived in harmony with nature

It is widely believed that early peoples, and people outside civilized societies today, lived in harmony with the environment, harvesting only enough for their needs and husbanding their resources.

It is true that in times past, mankind as whole did less damage to the environment in a year than we do today, as there were far fewer people around. However, each person probably individually did more damage per year. Over the centuries they brought about enormous changes to the landscape and to the flora and fauna. Here are a few examples.

- The English countryside is man-made and bears little or no resemblance to that which awaited the first settlers. The hills were then covered with scrub, and the valleys with forest.

- An African family uses 5 times more energy to cook their evening meal than a European one, and collecting that fuel, usually from forests, does more damage to the environment than cooking with gas or electricity.[1]
- The Mayan civilization of Central America, which flourished from 500 B.C. to 900 A.D., was responsible for removing most of the forests in the areas they occupied. Many reptiles and amphibians became extinct, and others were restricted to limited islands within the forest. The forests have recovered but not the animals.[2]
- A wooden canoe was displayed at an exhibition. "Wooden canoes," the notice said, "operate in harmony with the environment and are nonpolluting." However, a picture of the process of construction showed the Indians burning down large tracts of forest to obtain the correct timber and leaving most of the wood to rot.[3]

*References*

1. Baker, J. 1994. *Royal Society of Arts J* 142(5453):47.
2. Lee, J.C. 1990. *Natural History* Jan, p. 45.
3. Borley, N. 1986. *The Innocent Anthropologist,* p. 66. London: Penguin Books.

# MYTH E15

### Entropy is increasing all the time

We are often told that entropy, a measure of the degree of disorder in the universe, is increasing all the time as the universe "runs down." Erwin Schrödinger suggested that living organisms can reduce entropy, and James Lovelock[1] has further developed the idea. He suggests that the atmosphere on a planet not in chemical equilibrium, i.e., that contains gases that react with one another, indicates the presence of life. Thus the earth's atmosphere contains oxygen and methane, which react in sunlight; something must be producing one or both to replace them as they react. In contrast, the atmospheres on Mars and Venus are close to chemical equilibrium, indicating that they are lifeless (or that life is so limited as to be insignificant in its effects).

Another example is that astronomers tell us that the sun is about a third hotter now than when life started 3–4 billion years ago. Some self-regulating mechanism must be at work, or we would now be boiling. A possible mechanism is the production of dimethyl sulfide by algae in the oceans. Dimethyl sulfide forms nuclei in the atmosphere around which cloud droplets form.

*Reference*

1. West, R.E. 1989. Interview with J. Lovelock. *Orion Nature Quarterly* 8(1):58.

## MYTH E16

### Dioxin is the deadliest known poison

Whether or not dioxin is the deadliest known poison[1,2] depends on how we define "deadliest." Are we talking about the hazard, that is, the capacity to kill, or the risk, that is, the probability that people will actually be killed? Before we answer this question, we should be clear in our definition of dioxin.

The dioxins are a group of 210 compounds made by replacing one or more of the hydrogen atoms in dibenzodioxin or dibenzofuran by chlorine atoms. The most toxic of all is 2,3,7,8-tetrachloro-dibenzodioxin (2,3,7,8-TCDD). Seventeen of the others are also toxic, most to a much smaller extent, and the remaining 192 are nontoxic. Some of the literature on dioxins does not make it clear whether it is referring to 2,3,7,8-TCDD or to the group as a whole, and at one time it was not possible for analysts to distinguish between them. Today it is possible to make this distinction and to detect as little as one part per trillion, that is, 1 gram in a million tons. This is one reason for public concern about dioxins (and other toxins). When we are told that none has been detected, we relax. When analysts improve their methods and now detect even trace amounts, we start

dibenzo-dioxin                    dibenzo-furan

to worry, even though the amount is equivalent to only a few drops in a swimming pool.

Traces of 2,3,7,8-TCDD are formed during the manufacture of the herbicide 3,4,5-trichlorophenoxyacetic acid (3,4,5-T), and larger quantities are formed if the reaction is allowed to run away, as in the accident at Seveso in 1976.[3]

The 2,3,7,8-TCDD is certainly very toxic to some animals. The $LD_{50}$ (the amount that will kill half of those exposed) is 1 µg/kg of body weight for guinea pigs and 5 mg/kg of body weight for hamsters. For humans, the figure is not known but is certainly higher, perhaps very much higher. (For comparison, the $LD_{50}$ for botulism toxin is 10 µg/kg.) The 2,3,7,8-TCDD does, however, produce an unpleasant, but curable, skin disease called chloracne, a severe form of blackheads. All known cases have occurred to people heavily exposed as the result of chemical plant accidents, but no definite long-term effects have been observed. A study of 673 men heavily exposed following an accident in a chemical plant in 1968 found no overall excess of cancer; excesses of cerebrovascular disease and movement accidents may have been due to chance or to nonoccupational causes.[4,5]

The only deaths definitely due to dioxin occurred to four men who died after cleaning up the spillage from a runaway reaction in a chemical plant in the Netherlands in 1963.

Dioxins are produced during the combustion of many substances, including wood and municipal waste. Dioxins, probably produced by forest fires, have been detected in sediments 8000 years old. Humans have been exposed to dioxin from wood fires and cooking for hundreds of thousands of years and, compared with guinea pigs and hamsters, may have developed immunity to low concentrations.

Discharges of dioxins from the burning of waste, including chemical waste, and from the operation of chemical plants have been greatly reduced in recent years, and there is no evidence that present levels present any risk. To quote Christoffer Rappe, "More people make their living from dioxins than ever suffer from them."[1] Rappe is not just referring to the chemists who analyze dioxins but to the lawyers and doctors who deal with people who fear they may have been affected by dioxins.

To conclude, the hazards of dioxin may be high, but the risks are low.

*References*

1. Emsley, J. 1994. *The Consumers Good Chemical Guide,* Chap. 7. New York and Oxford, UK: Spektrum.
2. Whelan, E.M. 1993. *Toxic Terror,* Chap. 9. Buffalo, NY: Prometheus Books.
3. Kletz, T.A. 1994. *Learning from Accidents,* Chap. 9, 2nd ed. London: Butterworth-Heinemann.
4. Health and Safety Executive. 1994. *Mortality Study of Workers Employed at a Factory Manufacturing 2,4,5-Trichlorophenol.* London: Health and Safety Executive.
5. Marshall, V.C. *1995. Environmental Protection Bull* 35:17.

# 4

# Myths in Other Walks of Life

To purchase a clear and warrantable body of truth, we must forget and part with much we know.

Sir Thomas Browne

In all walks of life, we can find deeply ingrained beliefs that are not wholly true and that persist despite the lack of supporting evidence. I hope this book will encourage readers to look critically at much of the information they come across in newspapers, TV and radio programs, and elsewhere, and not take all they see or hear for granted.

One of the first writers to draw attention to myths was Sir Thomas Browne (1605–1682) who wrote in his Pseudoxia Epidemica, or Enquiries Into Very Many Received Tenets, and Commonly Presumed Truths (1642),[1] the quotation that introduces this part of the book. One of the myths he debunks is that "a man hath one rib less than a woman." This myth owes its origin to the Biblical story that Eve was made from one of Adam's ribs. Stephen Gould reports that during a TV call-in show, a young woman quoted this "well-known fact" as proof that the Bible was right and evolution wrong.[1] (Sir Thomas Browne also believed that the Bible was never wrong but said that, as one-armed men do not have one-armed children, a missing rib would not be passed on to offspring.) This example shows how people can believe in a myth that is easily disproved, in this case by examining skeletons.

While most myths are long-established, others are recent and some have been deliberately created. Eric Hobsbawn and Terence Ranger describe many of these in their book *The Invention of Tradition.*[2] For example, the Scottish kilt was invented by an Englishman in the 18th century, and the tartan patterns, different for each clan, were invented by 19th-century manufacturers of woollen cloth. The belief that the tartans preserve an old tradition is "an hallucination sustained by economic interest."[2]

The ceremonial opening of the UK Parliament by the monarch dates, in its present form, to the beginning of the 20th century. For the previous 40 years, Queen Victoria had never bothered to attend what was considered a formality.

In a fascinating book,[3] William Stafford has shown that there is no trustworthy evidence for most of the stories told about Mozart. One writer has copied from another, embellishing on the way, and repetition has given the stories an aura of truth.

Other examples of myths are the many folk remedies for the cure of illness. Some, of course, are effective, though unlike modern drugs, they have never been systematically tested for harmful side effects. Others are totally ineffective.

Turning to modern times, here is a summary of an article on six "myths that can drown you."[4]

1. A drowning victim will wave frantically while yelling for help
    Nonswimmers do try to attract attention by waving their arms; as a result, they find it difficult to keep their heads above water, and yelling is difficult or impossible.
2. Wait at least 30 minutes after eating before swimming
    Overeating before any exercise is dangerous, but a modest meal before swimming fuels the body and helps one keep warm.
3. If your boat capsizes, leave it and swim to the nearest shore
    The shore looks closer than it really is. Most boats do not sink when overturned, so by staying with the boat, you will have something to hold onto, and rescuers are more likely to see you.
4. A drunken person will sober up immediately when he hits the water

Only time will sober a person. Alcohol affects muscles and the brain, so swimming is more difficult when drunk.

5. If you fall into the water with your clothes on, take them off, so they will not weigh you down

Water in your clothes will not increase your weight. Air trapped in your clothes may help you to float, and the clothes will keep you warmer.

6. Once under water, you have only 4 minutes to live

This is not true in cold water (below 20°C), which slows down metabolism and reduces the demand for oxygen.

*References*

1. Quoted by Gould, S.J. 1995. *Natural History* June, p. 6.
2. Hobsbawn, E., and Ranger, T. 1983. *The Invention of Tradition.* Cambridge, England: Cambridge University Press.
3. Stafford, W. 1991. *The Mozart Myths.* Stanford, CA: Stanford University Press; and London: Macmillan.
4. Smith, D.S. 1986. *Family Safety and Health.* Chicago, IL: National Safety Council.

# Afterthoughts

. . . making the truth believable is much more difficult than making fiction plausible.

Daniel Easterman, New Jerusalems

Do not consider a thing as proof because you find it written in books . . . They are utter fools who accept a thing as convincing proof simply because it is in writing.

A truth, once established by proof, neither gains force by the assent of all scholars, nor loses certainty because of the general dissent.

Moses Maimonides (1135-1204)

Do not believe what your teacher tells you merely out of respect for your teacher. But whatsoever, after due examination and analysis, you find to be conducive to the good, the benefit, the welfare of all beings—that doctrine believe and cling to, and take it as your guide.

Gautama Siddhartha (c. 563-c. 483 BC)

# Glossary

Different words are used, in different countries, to describe the same job or piece of equipment. Some of the principal differences between the United States and the United Kingdom are listed here. Within each country, however, there are differences between companies.

Management Terms

| Job | United States | United Kingdom |
|---|---|---|
| Operator of plant | Operator | Process worker |
| Operator in charge of others | Lead operator | Chargehand, assistant foreman, or junior supervisor |
| Highest level normally reached by promotion from operator | Foreman | Foreman or Supervisor |
| First level of professional management (usually in charge of a single unit) | Supervisor | Plant manager |
| Second level of professional management | Superintendent | Section manager |
| Senior manager in charge of site containing many units | Plant manager | Works manager |
| Plant personnel | Craftsman or mechanic | Fitter, electrician, etc. |

The different meanings of the terms "supervisor" and "plant manager" in the United States and the United Kingdom should be noted.

Certain items of plant equipment have different names in the two countries. Some common examples are as follows.

Chemical Engineering Terms

| United States | United Kingdom |
|---|---|
| Accumulator | Reflux drum |
| Agitator | Mixer or stirrer |
| Air masks | Breathing apparatus (BA) |
| Blind | Slip-plate |
| Carrier | Refrigeration plant |
| Cascading effects | Knock-on (or domino) effects |
| Check valve | Nonreturn valve |
| Clogged (of filter) | Blinded |
| Consensus standard | Code of practice |
| Conservation vent | Pressure/vacuum valve |
| Dike, berm | Bund |
| Discharge valve | Delivery valve |
| Division (in electrical area classification) | Zone |
| Downspout | Downcomer |
| Expansion joint | Bellows |
| Explosion proof | Flameproof |
| Faucet | Tap |
| Fiberglass reinforced plastic (FRP) | Glass reinforced plastic (GRP) |
| Figure-8 plate | Spectacle plate |
| Flame arrestor | Flame trap |
| Flashlight | Torch |
| Fractionation | Distillation |
| Gasoline | Petrol |
| Gauging (of tanks) | Dipping |
| Generator | Dynamo or alternator |
| Ground | Earth |
| Horizontal cylindrical tank | Bullet |
| Hydro (Canada) | Electricity |
| Install | Fit |
| Insulation | Lagging |
| Interlock* | Trip* |
| Inventory | Stock |
| Lift-truck | Fork lift truck |
| Loading rack | Gantry |
| Manway | Manhole |
| Mill water | Cooling water |
| Nozzle | Branch |
| OSHA (Occupational Health and Safety Administration) | Health and Safety Executive |
| Pedestal, pier | Plinth |
| Pipe diameter (internal) | Pipe bore |

Chemical Engineering Terms (*Continued*)

| United States | United Kingdom |
|---|---|
| Pipe rack | Pipe bridge |
| Plugged | Choked |
| Rent | Hire |
| Rupture disc or frangible | Bursting disc |
| Scrutinize | Vet |
| Seized (of a valve) | Stuck shut |
| Shutdown | Permanent shutdown |
| Sieve tray | Perforated plate |
| Siphon tube | Dip tube |
| Spade | Slip-plate |
| Sparger or sparge pump | Spray nozzle |
| Spigot | Tap |
| Spool piece | Bobbin piece |
| Stack | Chimney |
| Stator | Armature |
| Tank car | Rail tanker or rail tank wagon |
| Tank truck | Road tanker or road tank wagon |
| Torch | Cutting or welding torch |
| Tower | Column |
| Tow motor | Fork lift truck |
| Tray | Plate |
| Turnaround | Shutdown |
| Utility hole | Manhole |
| Valve cheater | Wheel dog |
| Water seal | Lute |
| Wrench | Spanner |
| C-wrench | Adjustable spanner |
| Written note | Chit |
| $M | Thousand dollars |
| $MM | $M or million dollars |
| STP | 60°F, 1 atmosphere |
| 32°F, 1 atmosphere | STP |
| NTP | 32°F, 1 atmosphere |

*In the United Kingdom, "interlock" is used to describe a device that prevents someone from opening one valve while another is open (or closed). "Trip" describes an automatic device that closes (or opens) a valve when a temperature, pressure, flow, etc., reaches a preset value.

# Index